Practical Guide to Chemometrics

Practical Guide to Chemometrics

Stephen John Haswell
University of Hull
Hull, England

Marcel Dekker, Inc.   New York • Basel • Hong Kong

Library of Congress Cataloging-in-Publication Data

Practical guide to chemometrics / [edited by] Stephen J. Haswell.
    p.   cm.
  Includes bibliographical references and index.
  ISBN 0-8247-8597-5 (acid-free paper)
  1. Chemistry, Analytic--Statistical methods. 2. Chemistry, Analytic--Mathematics. I. Haswell, S. J. (Stephen John)
QD75.4.S8P73 1992
543'.0072--dc20   91-41642
                                                              CIP

This book is printed on acid-free paper.

Copyright © 1992 by MARCEL DEKKER, INC. All Rights Reserved

Neither this book nor any part may be reproduced or transmitted in any form or by any means, electronic or mechanical, including photocopying, microfilming, and recording, or by any information storage and retrieval system, without permission in writing from the publisher.

MARCEL DEKKER, INC.
270 Madison Avenue, New York, New York 10016

Current printing (last digit):
10 9 8 7 6 5 4 3 2 1

PRINTED IN THE UNITED STATES OF AMERICA

Preface

This book has been produced as a guide and general text for the scientist interested in finding out more about how chemometrics can improve the practice of their science.

Scientists who are involved with chemical analysis as part of their work, in particular the analytical chemist, will find this book a practical guide to the subject of chemometrics. This developing area offers the analyst an opportunity to critically look at both the quantity and quality of experimental data. In short, chemometrics is a powerful tool that offers the analyst access to a greater amount of more reliable analytical information using existing instrumentation. For many, however, the problems of where to start and which particular techniques to select have limited the use of chemometrics. We hope this text will solve such problems.

The central theme of chemometrics is the laboratory instrument, and the chapters of this book discuss and illustrate, through worked examples, how computational methods can enhance experimental data. The text restricts itself to the mathematics considered necessary for the basic understanding of the various techniques covered, as in practice most analysts today interact with computers and not calculators.

The early chapters examine the ways in which instrumental data and calibration can be evaluated and how data can be assessed for analytical quality. Getting the best out of your instrument is the focus of the optimization chapter, while getting the most out of your signal is the theme of the chapter on signal processing. As analysts we are familiar with the basic principles of our instrumentation; it is therefore fitting that one chapter is dedicated to addressing the role of the microprocessor in instrumental control and signal management. Once again this chapter has tried to keep the description of the microprocessor as lucid as possible to give the reader a feel for the role of this device in signal processing. The final chapter considers some examples of areas of current and future applications of chemometrics.

The authors of the individual chapters are all practicing chemometricians, which brings an applied quality to the subjects discussed. Above all, the aim of each has been to produce a book for the practicing scientist that forms an introduction to chemometrics and the techniques encompassed by the discipline, emphasizing the practical nature of the subject and its role in the modern laboratory. We hope the book will encourage you to get the best out of chemometrics.

Stephen John Haswell

Contents

Preface iii

Contributors ix

Chapter 1 Introduction to Chemometrics—*Stephen John Haswell* 1

1.1 Getting the Most Out of the Book 2
1.2 General Reading on Chemometrics 2

Chapter 2 Statistical Evaluation of Data—*Stephen John Haswell* 5

2.1 Sources of Error 7
2.2 Precision and Accuracy 8
2.3 Significance Testing—the Student t and F Tests 15
2.4 Analysis of Variance 26
2.5 Outliers 32
2.6 Testing for Normality of Distribution 34
2.7 Comparison of Two Methods by Least-Squares Fitting 36
    References 37

Chapter 3 Exploratory, Robust, and Nonparametric Data Analysis—
John Michael Thompson 39

3.1 Limitations of Parametric Data Analysis 43
3.2 Nonparametric Methods 49
3.3 Exploratory Data Analysis 62
3.4 Robust Methods 85
   References 91
   Software and Bibliography 96

Chapter 4 Calibration—*John H. Kalivas* 99

4.1 Sampling Theory 100
4.2 Least Squares and Linear Calibration 111
4.3 Confidence Intervals 124
4.4 Nonlinear Calibration 128
4.5 Sensitivity and Limits of Detection 130
4.6 Interference Effects 137
4.7 Standard Addition 139
4.8 Solutions to Problems 144
   References 147

Chapter 5 Nonlinear Regression Analysis—*James F. Rusling* 151

5.1 Linear and Nonlinear Models 152
5.2 Convergence Methods 154
5.3 Practical Model Building 162
5.4 Applications 168
5.5 Problems and Solutions 171
   References 178

Chapter 6 Experimental Design and Optimization—*M. J. Adams* ... 181

6.1 Randomized Blocks 184
6.2 Latin Squares 187
6.3 Factorial Designs 189
6.4 Response Surfaces and Optimization 198
   References 209

Chapter 7 Signal Processing and Data Analysis—*Barry K. Lavine* ... 211

7.1 Data Representation 212
7.2 Data Preprocessing 213

Contents vii

7.3 Mapping and Display 213
7.4 Clustering.. 216
7.5 Classification... 219
7.6 Applications of Pattern Recognition Techniques 227
    References 236

Chapter 8 Signal Processing and Data Enhancement—*Steven D.
Brown* 239

8.1 Noise Removal and the Problem of Prior Information 240
8.2 Frequency Domain Signal Processing 242
8.3 Frequency Domain Smoothing.. 251
8.4 Time-Domain Filtering and Smoothing 255
8.5 Resolution Enhancement in the Time and Frequency Domains 262
8.6 Summary... 267
8.7 Solutions to Problems... 267
    References 268

Chapter 9 The Role of the Microcomputer—*K. L. Ratzlaff and
Eugene H. Ratzlaff*... 271

9.1 The Microprocessor 271
9.2 Algorithm Design... 285
9.3 Software and Programming 286
9.4 Laboratory Interfaces... 292
9.5 Laboratory Information Management Systems 303
9.6 Artificial Intelligence Activities with Microcomputers................ 305
    References 307

Chapter 10 Trends in Chemometrics—*Steven D. Brown* 309

10.1 Chemometrics and Information Theory 310
10.2 The Future of Chemometrics 314
10.3 Solutions to Problems 318
     References 319

Index 321

Contributors

M. J. Adams   School of Applied Science, Wolverhampton Polytechnic, Wolverhampton, England

Steven D. Brown   Department of Chemistry, University of Delaware, Newark, Delaware

Stephen John Haswell   School of Chemistry, University of Hull, Hull, England

John H. Kalivas   Department of Chemistry, Idaho State University, Pocatello, Idaho

Barry K. Lavine   Department of Chemistry, Clarkson University, Potsdam, New York

Eugene H. Ratzlaff   IBM Research Division, T. J. Watson Research Center, Yorktown Heights, New York

K. L. Ratzlaff   Department of Chemistry, Instrumentation Design Laboratory, The University of Kansas, Lawrence, Kansas

James F. Rusling   Department of Chemistry, The University of Connecticut, Storrs, Connecticut

John Michael Thompson   Biomedical Engineering and Medical Physics Department, University of Keele Hospital Centre, Stoke on Trent, England

1
Introduction to Chemometrics

Stephen John Haswell *University of Hull, Hull, England*

The term *chemometrics* was coined in 1971 to describe the growing use of mathematical, statistical, and other logic-based methods in the field of chemistry and in particular in analytical chemistry. The application of chemometrics has found considerable success in three general areas: (1) the calibration, validation, and significance of analytical measurement; (2) the optimization of chemical measurement and experimental procedures; and (3) the extraction of the maximum chemical information from analytical data. It is in the latter area that chemometrics has become a very powerful technique offering multivariate data reduction procedures that are now yielding chemical information previously not available to the analyst from data. Many of the so-called chemometric techniques obviously existed prior to the 1970s, but the philosophy of the techniques as a discipline of chemistry has seen considerable growth as a subject over the past decade. It is perhaps not surprising that this growing interest has been supported by the increasing amount of analytical data that can now be obtained for modern instrumentation often linked to a Laboratory Information Management System (LIMS).

Many scientists whose work relies on the production of chemical data are now aware of the importance of obtaining reliable data and can see the value of instrumental optimization, signal processing, and data integration methods. In general, many of the techniques available are relatively simple to use, but where to start and what to do are often difficult questions to answer. Many of the existing texts on the subject focus on the algebraic basis of the chemometric methods, and the analyst has found it hard to break into the subject area. The authors of this book have taken some care to produce a book that will introduce the more popular chemometric methods and guide the interested scientist through them so that he or she will come to use these valuable techniques. The application of these techniques is not without problems, and this book directs the reader to potential problem areas in attempting to process and evaluate data.

1.1 GETTING THE MOST OUT OF THE BOOK

Each of the chapters that follow is self-contained; and together they cover the main areas of chemometrics. An introduction at the start of each chapter gives the reader an indication of the applicability of the subject area, and authors have used selected examples to illustrate their text. For the reader who is a little uncertain about where to start, the following brief summary may be of assistance.

Chapters 2 and 3 examine methods of treating experimental data to evaluate them for quality, and many of these methods together with those of Chapters 4 and 5 will be of interest to scientists involved in data quality assurance and control. Chapter 6 gives details of methods that can be used in the design of efficient experimental procedures and will be of interest to those interested in the development of new analytical methodology or wishing to optimize experimental methods. This aspect of the subject has some relevance also to the material covered in Chapter 8. The growing area of multicomponent data analysis is considered in Chapter 7, which addresses the valuable role chemometrics has in extracting information. Chapter 7 will be of considerable interest to workers attempting to handle and interpret large data sets. The treatment of our analytical signals is a relatively simple process with today's process-controlled instrumentation, but what it all means is the subject of Chapter 8, which the discerning analyst concerned about his or her data will find of interest. The computer is now commonplace in almost every analytical laboratory, but are we getting the best out of our micro? Chapter 9 deals with this subject in a very lucid way. Finally, in Chapter 10 we conclude with a consideration of some of the directions in which chemometrics may be moving and future areas of application.

1.2 GENERAL READING ON CHEMOMETRICS

A growing number of books are available on chemometrics, many of a specialized nature. A brief summary of the more general texts are given here as guidance for the reader. However, each chapter of this book has its own list of selected references.

INTRODUCTION TO CHEMOMETRICS 3

D. L. Massart, B. G. M. Vandeginste, S.N. Deming, Y. Michotte, and L. Kaufman, *Chemometrics: A Textbook*. Elsevier, Amsterdam, 1988.

B. R. Kowalski, *Chemometrics: Theory and Applications*, ACS, Washington, D.C., 1977.

G. Kateman and F. W. Piipers, *Quality Control in Analytical Chemistry*, Wiley, New York, 1981.

M. A. Sharaf, D. L. Illman, and B. R. Kowalski, *Chemometrics*, Wiley, New York, 1986.

R. G. Brereton, *Chemometrics: Applications of Mathematics and Statistics to Laboratory Systems*, Ellis Horwood, Chichester, 1990.

J. C. Miller and J. N. Miller, *Statistics for Analytical Chemistry*, 2nd ed., Ellis Horwood, Chichester, 1988.

Two very good journals also exist on the subject.

Journal of Chemometrics and Intelligent Laboratory Systems (Elsevier)—Good for conference information; has a tutorial approach and is not too mathematically heavy.

Journal of Chemometrics (Wiley)—Good for fundamental papers and applications of advanced algorithms.

Papers on chemometrics can also be found in many of the more general analytical journals.

2
Statistical Evaluation of Data

Stephen John Haswell   *University of Hull, Hull, England*

This chapter focuses on the reporting and evaluation of results based upon statistical techniques. It is by using such techniques that a mathematical measure of the significance of data is achieved. For example, we can qualify a question such as Is the concentration of aflatoxin in a food sample significantly high to cause a toxic effect? Such tests are very important in analytical chemistry, especially when some level of confidence in our quantification is required. This is particularly relevant in the estimation of trace levels of a determinant where the chemist has to judge whether the levels of an analyte are low or whether it is actually present. What is implicit in our use of statistics is that we are acknowledging the errors that occur in our analysis. It is clear, therefore, that to obtain quantitative results we must make some measure of the errors inherent in our determinations.

The use of statistical methods to gain some level of confidence in our data is not without problems, however, and we should not use these methods blindly. The following sections serve to illustrate some of the more common techniques used by the analytical chemist; this chapter is in no way an exhaustive review of methods. We will see that the concept of accepting or rejecting the hypothesis that sets of data are the same or, to be more statistically correct, that there is no difference between the distribution of data populations at some predictable level of significance forms the basis of running of those tests. The difference between the characteristics of the sample and those of the population can lead to an erroneous conclusion, so it is important to use the tests described in an intelligent way if we are to have any confidence in the interpretations we make. The following criteria may act as a guide to the decision making required in performing a statistical evaluation of data. The elaboration of a test consists of a number of steps:

1. Clearly identify the question you wish to answer.
2. Select the appropriate test, and look carefully at similar tests to see which suits your circumstance best (do not do them all in the hope that one will work).
3. Decide the level of significance you wish your test to be at; this is usually the 5% level (i.e., 1 in 20), but you may wish to consider the 1% or even 0.01% level.
4. You will probably adopt the null hypothesis, which assumes that there is no difference between the data sets being tested at the level of significance you have selected (innocent unless proved quilty). There is also the alternative hypothesis available, which assumes that there is a difference unless the presence of no difference is indicated by the test. The first method—the null hypothesis, which you will meet throughout this chapter—will probably be your first choice.
5. Perform the test statistic.
6. Compare the test statistics with the tabulated critical values obtained from statistical theory (the yardstick). Remember to allow for the number of degrees of freedom or nonexclusiveness in the data set.
7. The final, and most important, step is acceptance or rejection of the null hypothesis. In general, if your value is lower than the theoretical or critical value in the tables for a given level of significance the null hypothesis holds true—there is no difference in the data sets. On the other hand, if the value is greater, than we reject the hypothesis and conclude that there is a statistical difference. For the Mann–Whitney test the null hypothesis is accepted when the test value is higher than the critical value. At this final stage, care in adopting the hypothesis is important, and in this chapter we shall discuss potential sources of error in our interpretation.

It is the aim of this chapter to guide the reader to commonly used significance tests. Each section will describe these tests and give an overview of how they can be used to assist the analytical chemist in his or her work. References are indicated throughout the chapter should more detailed information be required.

2.1 SOURCES OF ERROR

In the experimental work we all perform as analytical chemists there may occur three main types of errors: gross, systematic, and random errors.

The first of the three, the gross error, is probably the most obvious and can be attributed to problems such as instrumental breakdown or severe reagent or sample contamination. Errors of this type represent such serious distortions of the data that the experimental information should simply be rejected. Systematic and random errors, on the other hand, are problems we are going to have to manage every day in our experimental results.

Systematic errors, as the name implies, occur as a result of some bias in the system (instrumental or procedural), such as an incorrectly calibrated instrument or the use of wrong or inaccurate volumetric glassware. Systematic errors can further be classified into two types; constant and proportional. Those of the constant type are found to be independent of the analyte value or concentration and may result from poor blank values or the presence of a component in the analyte matrix that inhibits measurement or reacts with the analyte to interfere with signal production. Clearly, if the matrix effect becomes too severe it may constitute a gross error. Proportional errors will vary with analyte concentration and are commonly associated with calibration errors in which the slope of the calibration line is different from the slope of analyte calibration in a real matrix. Standard addition techniques are often of great value in identifying this form of error [1]. It is through an estimation and experimental reduction of systematic errors that one can determine the accuracy or closeness of a determination to the true value for a set of measured values.

The final type of error, random error, arises from situations over which we have little or no control, such as electronic noise in a transducer, which leads to uncontrolled random variations in our data approximating to an average value. It is these random errors that affect the precision or reproducibility of our experimental results. When the random error is small our results are said to have a good or high level of precision. As the precision of a method is determined from replicate measurements taken more or less at the same time, an additional comparative measurement—repeatability—may also be of use to the analyst. This measurement can be used to assess the day-to-day or month-to-month variability in experimental data.

So we see that for any set of data the values we obtain are subject to errors. It is worth noting that the major source of error in any analytical work must be the sampling, which is very much a subject in its own right and is not included in the scope of this book. The reader interested in sampling strategies and methods is advised to refer to texts written specifically on the subject [2]. In general, however, the techniques available for acquiring representative samples from a population should obviously reflect the sample population or system under investigation, and such techniques are almost as varied as the types of samples analyzed. The major underlying factor that leads to gross errors, often on the order of 80–90% of total error, is the lack of homogeneity in samples, be they water bodies or solids. It is clearly the responsibility of the laboratory-based analyst to take considerable care in ensuring that any sample presented for analysis is made as homogeneous and representative as possible. It is becoming increasingly important for analysts to develop a multidisciplinary approach to their analysis so that the interface between data acquisition (which over the years analysts have become good at) and data interpretation (until now somebody else's problem) can become more meaningful.

We have seen from this introduction to errors that our total analytical error, excluding that associated with sampling, can be summarized as

Total analytical error = random error + systematic error

It is therefore the responsibility of the analyst to develop good practice in evaluating analytical error and so obtain the best quality results possible. As a general guide, it is best to aim for good precision first and then assess the accuracy of one's results. It is not reasonable to expect accurate results unless the precision is good. The remaining sections of this chapter address techniques that will assist the analyst in determining errors in their analytical methodology.

2.2 PRECISION AND ACCURACY

Precision

The first step in evaluating the quality of our data is to look at how well we can repeat the analyte determination for a given sample. Various bodies have attempted to define precision, including the International Union of Pure and Applied Chemistry (IUPAC), who define precision as "relating to the variations between variates, ie scatter between variates" [3]. A more common definition is "Precision refers to the reproducibility of measurement within a set, that is, to the scatter or dispersion of a set about its central value" [4]. In this case the word *set* is used to describe the number of replicate measurements. Before we become waylaid by such definitions, let us remind ourselves that what we are attempting to do is measure the random error in our analysis. To make this assessment we take our first steps into statistics, in that we assume that our data follow a predetermined pattern of distribution about some central or mean value. We base our evaluation on the premise that our data follow a normal or Gaussian distribution, which means that there is a well-defined and well-understood mathematical distribution to our data set. By employing such a distribution we can assess the random error in a data set by statistically testing our data using the distribution model.

STATISTICAL EVALUATION OF DATA

The concept of a data set following a mathematically described distribution forms the basis of most of the statistics described in this section. You may also be familiar with other distributions such as the binomial or poisson distribution, which are not dealt with in depth in this chapter. You may recall that data display a binomial distribution if the population consists of groups belonging to two mutually exclusive categories. The Poisson distribution also describes discrete events but with a continuous interval such as time. For example, radio immunoassays often follow a Poisson distribution; however, when n becomes >10 the distribution starts to tend to a normal distribution. There are numerous other distribution models and tests to estimate how well data fit a given distribution. The reader is directed to reference 5 for a more specialized treatment on the range of distributions available.

It is worthwhile at this point to briefly consider the normal distribution in some detail in order to see how we make practical use of this important distribution model. Take, for example, the measurement of 50 samples selected from, in theory, an infinite number of possible samples (e.g., water in a reservoir), the population.

If there were no systematic errors, then the mean of this population (the entire reservoir), denoted traditionally by μ, would be the true value of the analyte present (not an easy value to obtain in most cases). On a more practical level, the mena for our fifty 100-mL samples of reservoir water, $\bar{x}$, represents only an estimate of μ (the mean $\bar{x}$ is calculated from $\bar{x} = \Sigma x/n$, where Σx is the sum of all the measurements and n is the number of measurements). But if we consider for a moment the total population, which we assume is truly homogeneous, then if we were to determine all the samples possible (i.e., all the water in a reservoir taken as 100-mL samples) the data would naturally give a spread or scatter around μ due to random errors in the measurement. The spread of the data follows a well-defined pattern or distribution that we call the *normal distribution* (see Figure 1), the line of which is described by the equation

$$y = \frac{\exp[-(x-\mu)^2/2\sigma^2]}{\sigma\sqrt{2\pi}} \tag{1}$$

where μ is the true value, x is each measurement, and σ is the standard deviation.

It is not important to remember this equation, but to remember the concept—that the data spread described by the standard deviation (σ) is most relevant. The standard deviation of our sample (identified by the symbol S) is therefore only an estimate of the true standard deviation given for the infinite population by σ. It is common in statistics to use Greek symbols for the population statistic and arabic symbols for the sample population. The standard

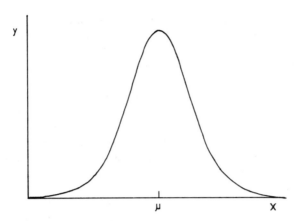

Figure 1 The normal distribution.

deviation for our 50 samples is therefore a measure of spread in the data (or random error) and is described by the formula

$$S = \sqrt{\Sigma_i\ (x_i - \bar{x})^2 / (n-1)} \qquad (2)$$

where the x_i are the values for the individual samples, $\bar{x}$ is the mean for the sample population, n is the number of samples, and Σ is a summation term.

So we see that the standard deviation is basically determined by polling together the differences between real values x and the mean $\bar{x}$. Most pocket calculators will give the standard deviation if the x values are entered; however, care should be taken to select the correct standard deviation. At this point you may be asking yourself two questions; first, What if my data are not normally distributed? and, second, Why do we complicate the issue by having different ways of calculating the standard deviation? The answers to both these questions relate to the size of the sample population we use. If we had the time and the money to perform more than 30 replicate determinations on each sample for every analyte, then the statistic would be simple.

The central limit theorem evoked by statisticians states that as the number of samples increase (≥ 30) the data will tend to a normal distribution (not surprising if you think about it). Briefly, the central limit theorem can be considered by the following:

$$y = x_1 + x_2 + \ldots + x_n \qquad (3)$$

where n is the number of independent variables x_i that have a mean μ and variance σ^2 ($i = 1, 2, \ldots, n$).

STATISTICAL EVALUATION OF DATA 11

For a large number of variables n, the distribution of y is approximately normal, with mean $\Sigma\mu$ and variance $\Sigma\sigma^2$, despite whatever the distributions of the independent variable x might be. This explains why, in chemistry, distributions of error are often a function of many component errors that can be approximated to a linear function of independently distributed component errors. The important consideration is, however, that the component errors have the same order of magnitude and that no significant single source of error dominates all the others (this is often not the case in chemistry). For our purposes the important application of the central limit theorem is the way it describes the data distribution around the mean. Returning to our reservoir example, if all the water was sampled as 100-mL aliquots and we know that the true mean is μ with a standard deviation of δ, then the distribution of the sample mean will have a mean of $\mu_x = \mu$ and a variance of $\sigma_{\bar{x}}^2 = \sigma^{2/n}$. Thus the sample distribution will be normally distributed when the parent population is normally distributed or will appear so when $n \geq 30$ regardless of the shape of the parent population. The problem that arises, however, is that as anlysts we are happy to do duplicate anlyses but it gets a little more awkward when we are required to do 10 let alone 30 replicate analyses on each analyte. So for small numbers of replicate values (≤ 30), rather than using n in Eq. (2) we use $n-1$ as the denominator, where the -1 is known as a *degree of freedom*. It is recommended that a minimum of six replicate samples be used to obtain a reliable standard deviation, which in Eq. (2) will correspond to five values (i.e., $n-1 \equiv 6-1 = 5$).

Degrees of freedom often confuse the analyst, and it is not always clear why we complicate the issue with them, but there is a good statistical reason. The degree of freedom used with small data sets accounts for the quite high likelihood that a value in the deviation may not be truly independent. To put it another way, the term *degrees of freedom* refers to the number of mutually independent deviations $(x-\bar{x})$ that are used in calculating the standard deviation. In this case the number is $n-1$, because when $n-1$ deviations are known the last value can be deduced by using the result $\Sigma(x_i-\bar{x}) = 0$.

The standard deviation (S) can be expressed in various ways. For example, the square of S is known as the *variance* (S^2), while the relative standard deviation (RSD) or coefficient of variation (CV), popular terms with analytical chemists, is obtained from $100 \times S/\bar{x}$ and is expressed as percent relative error. The way of expressing the standard deviation is a choice for the analyst. If one chooses to use the standard deviation, then the units of measurement given to the mean are inherent in the expression of random error; however, the RSD is a convenient way to express the random error regardless of concentration or weight of the analyte measured. In this way the random errors for different concentrations and for different methods can be easily compared. So we see that by estimating the spread of our data around a mean value we can determine the

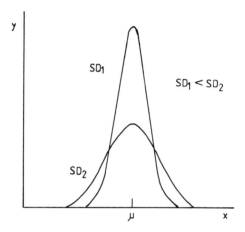

Figure 2 The normal distribution with the same mean but different standard deviations.

precise or random error in our determinations. The actual slope of the curve or its symmetry around the mean $\bar{x}$ will be a function of the standard deviation (see Figure 2). Regardless of the slope, detailed statistical analysis tells us that 68% of the data will lie within ±1 standard deviation S of the mean, approximately 95% of values will lie within ±2S of the mean, and approximately 99.7% of the values will lie within ±3S of the mean (Figure 3). So by selecting 1, 2, or 3 standard deviations around our mean, for any given data set we can estimate the amount of spread (or random error) in our data that fit into the 68%, 95%, or 99.7% window.

It is worth recalling at this point that we are in reality using our sample mean $\bar{x}$ to estimate the true value μ for the analyte in question. Since we know that our data are spread around the true value, a factor that depends on precision, it is most unlikely that the sample mean $\bar{x}$ we determine is exactly equal to the true value (a point we will return to in the next section). For this reason it may be better to express values as a range within which we are almost certain the true value lies. The width of this range depends on two factors: (1) the precision of the individual measurements and (2) the number of measurements made. In the example given below, it can be seen that when the precision is determined on the first six values rather than on all 10, then the spread is greater; however, if we were to increase the number of values to say 20 or 40, then not only would we get better precision (a point seen earlier) but the mean $\bar{x}$ might become closer to the true value μ. So if we subdivide a data set to obtain a number of means, rather than one mean, in order to estimate more closely the true value μ, we will get a distribution of mean values rather than individual data points. This produces for us a deviation known as the *standard error of the*

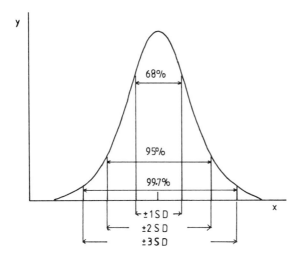

Figure 3 The normal distribution showing that approximately 68% of values lie within ±1 standard deviation, 95% lie within ±2 standard deviations, and 99.7% lie within ±3 standard deviations.

mean (SEM), which relates to the standard deviation of the population according to

$$\text{SEM} = \sigma/\sqrt{n} \qquad (4)$$

As we expect, the larger n is, the smaller will be the spread or error about μ. The standard error of the mean is not widely used in evaluating analytical data, but the analyst should be aware of its existence and what it represents.

Returning to our standard deviation, we now see that we have a good tool for evaluating the preference of our analytical methodology with reference to random errors. To summarize, let us briefly look at one example. The two following data sets might be replicate pH values of an alkaline solution (set A) and an acidic solution (set B).

Set A: 11.2, 10.7, 10.9, 11.3, 11.5, 10.5, 10.8, 11.1, 11.2, 11.0
Set B: 5.2, 6.0, 5.2, 5.9, 6.1, 5.5, 5.8, 5.7, 5.7, 6.0

For set A,

	$n = 10$	$n = 6$
$\bar{x}$	11.0	11.0
S	0.301	0.381
RSD	2.7%	3.5%

and for set B,

	n = 10	n = 6
$\bar{x}$	5.6	5.6
S	0.341	0.403
RSD	6.0%	7.1

The precision of data set A is calculated to be better than that of data set B (2.7% RSD is more precise than 6.0% RSD), so we can conclude that data set A has less random error than data set B. Note how the precision deteriorates as the population becomes smaller ($n=6$), while the mean does not apparently change, a point discussed earlier in this section.

To conclude, it is always difficult to say what is good or bad precision, as it depends very much on the particular sample type and methodology adopted. In general, for a routine instrumental analytical method, an RSD of less than 2% is usually acceptable, with values greater than 5% indicating a problem with random noise errors. As a final point, the analyst is reminded that it is my view that when assessing the random errors for a given method, a minimum of six and preferably 10 replicate values should be taken.

Accuracy

Accuracy is the parameter we measure to estimate just how close our determinant value is to the true value. The odd thing about this determination, as we shall see, is that often the true value is not known and so the estimate itself is prone to errors. A formal definition of *accuracy* might be that it is the difference between the determined mean $\bar{x}$ for a set of results and the value that is accepted as the true or correct value for the analyte measured, denoted by μ_o. As we have seen from the previous section, whenever we make a determination x of an analyte the value obtained is likely to be different from the true value μ_o, which may or may not be known. This difference is known statistically as the error (e) of x. Thus

$$e = x - \mu_o \qquad (5)$$

However, if a large number of determinations are made, then the mean $\bar{x}$ will tend to be closer to the true value μ_o obtained from an infinite number of determinations. The absolute difference between μ derived from $\bar{x}$ and the true value μ_o is called the *systematic error* or *bias*; thus e becomes

$$e = \underbrace{x - \mu}_{\text{random error}} + \underbrace{\mu - \mu_o}_{\text{bias}}$$

What you have probably realized is that as $\bar{x}$ and hence μ are only estimates obtained by experimentation, there will be some error in the estimation of the systematic error or bias. Problems with the random error estimate have been discussed in an earlier section.

The types of bias that are common in analytical methodology are laboratory bias and method bias. *Laboratory bias* is something that occurs in specific laboratories and may result, for example, from an uncalibrated balance or contaminated still. This source of bias is easily shown up when results from interlaboratory studies are compared and statistically evaluated. *Method bias*, on the other hand, is not readily distinguishable between laboratories following a standard procedure, but can be identified when reference materials or characterized material or methods are used. Those of us involved in data quality assurance will become only too familiar with the importance of interlaboratory studies and the use of reference material to evaluate the accuracy of our analysis.

In the next section we will see the procedures that we can carry out to test our data for significant errors in precision and accuracy.

2.3 SIGNIFICANCE TESTING— THE STUDENT t AND F TESTS

It is now time to introduce a new meaning to the word *significance*. In a statistical sense we are able with some level of confidence to say whether or not there is a significant difference between, say, two sets of data. In this case, "significant" infers that a test has been carried out, and it would be incorrect to state when reporting results that no significant difference was observed if a suitable test was not carried out. The obligation lies with the analyst to evaluate the significance of the results and report them in a correct, unambiguous manner. So we should be using significance testing to evaluate the quality of our results by estimating systematic (accuracy) and random (precision) errors in our experimental data. In this section we will be looking at the various tests that have particular relevance to the analytical chemist.

The easiest way to estimate the accuracy of a method is to analyze a standard or reference material for which there is a known or true value μ for our analyte. Thus the difference between the experimentally determinant mean $\bar{x}$ and the true value μ will be due to both the method bias (systematic error) and random errors. As indicated, we employ significance testing to estimate the proportion of each type of error. It is usual to first estimate whether any deviation is due solely to random errors. This is achieved through the use of the Student t test, more commonly called just the t test, which will show if the experimental mean $\bar{x}$ does or does not significantly differ from the true value μ. In this case the deviation between the given and experimental value is considered to be due to random errors, and the method can be assessed for accuracy. If this assumption is not made, then the deviation becomes a measure of the systematic error or bias. This approach to accuracy testing is limited, however, to samples where reference materials are available. The more usual case is that no such material is present, so a reference method that is considered to have no laboratory bias is used to represent the true value. Note that the reference method only represents a true value and may itself be incorrect; however, the methodology does represent a procedural standard. The variance or data spread of, say, two comparative methods can be compared using the F test; in this case we are assuming that any differences in the random errors are present in replicate measurements. It is sometimes necessary to carry out an F test prior to a t test in order to establish that the random errors in comparable methods are not significantly different. As we will see later, this is important in establishing that there are no significant differences in the standard deviation distributions for the two comparable methods.

In the tests that follow we are simply comparing our data (summarized by a figure of merit) to a set of model statistical values that have been obtained mathematically from a characterized distribution. By measuring our test data against the theoretical value we can estimate whether or not there is a statistically significant difference between our data and the mathematical model. In other words, we test our experimental data against the statistically expected result based on a known distribution, usually our friend the normal distribution.

The main point to grasp about the tests is that they are based upon what is known as the null hypothesis. Using the null hypothesis, you basically assume that your experimental and test data are the same unless the test statistic indicates that they are different. If we assume that the null hypothesis is true—that there is no difference between the two sets of numbers—then statistical theory can be used to calculate the probability or chance that a difference between, say, the sample mean and a true value will occur. Therefore, if we find that our test value for our data is worse than the critical value given in the theoretical table (i.e., the test value is numerically greater than the critical value), then we conclude that there is a significant difference at, say, a 5% confidence level in our data. It is important to note, however, that if we accept the null hypothesis and conclude that there is no significant difference at say, the 5% level, we are not proving that the value is true but only qualifying that statistically it is not a false assumption at the level of confidence we have selected.

It is time now to look more closely at the mechanism of the t test and see the major ways in which it can be used. The basic idea of the test is that we determine a mean $\bar{x}$ for our sample, obtained from at least six replicate values, and test it against the true value μ or something we consider to be representative of a good result. Note that we are now operating with small data sets where $n \leq 30$, more typically $n = 5\text{-}10$; as n becomes ≥ 30 we move to a normal

STATISTICAL EVALUATION OF DATA

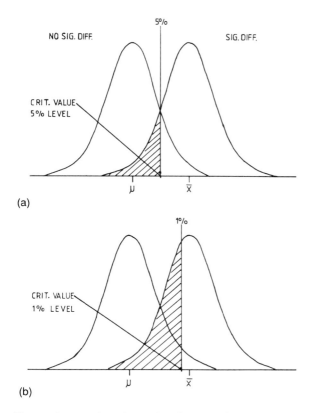

Figure 4 Overlap of two t distributions, showing variation in the critical values for the 5% (a) and 1% (b) significance levels.

distribution. We carry out the t test using distribution profiles similar to the normal distribution, so the t distribution, like the normal distribution, will be a function of n, and the greater n is, the closer our fit will become to the ideal distribution. The region of overlap in Figure 4 represents the area in which there are differences in the data. If we draw our line of significance at A (5% level), then if the test value for $\bar{x}$ lies to the left of the line we assume that there is no significant difference. By moving the line to the right—for example, to position B (1% level)—we increase the level of significance. In practice we use t tables, not distribution patterns, to make our assessment of whether there is or is not a significant difference in our comparative data (Table 1). In a moment we shall see how to calculate the t value for our data, which becomes our figure of merit, and how we test it against the critical or statistically determined value given in

Table 1 The t Distribution

Value of t for a confidence interval of	90%	95%	98%	99%
Critical value of $|t|$ for P values of	0.10	0.05	0.02	0.01
Number of degrees of freedom				
1	6.31	12.71	31.82	63.66
2	2.92	4.30	6.96	9.92
3	2.35	3.18	4.54	5.84
4	2.13	2.78	3.75	4.60
5	2.02	2.57	3.36	4.03
6	1.94	2.45	3.14	3.71
7	1.89	2.36	3.00	3.50
8	1.86	2.31	2.90	3.36
9	1.83	2.26	2.82	3.25
10	1.81	2.23	2.76	3.17
12	1.78	2.18	2.68	3.05
14	1.76	2.14	2.62	2.98
16	1.75	2.12	2.58	2.92
18	1.73	2.10	2.55	2.88
20	1.72	2.09	2.53	2.85
30	1.70	2.04	2.46	2.75
50	1.68	2.01	2.40	2.68
∞	1.64	1.96	2.33	2.58

The critical values of $|t|$ are appropriate for a two-tailed test. For a one-tailed test the value is taken from the column for twice the desired P value; e.g., for a one-tailed test, P = 0.05, 5 degrees of freedom, the critical value is read from the P = 0.10 column and is equal to 2.02.

Tabular extracts taken from *Elementary Statistics Tables* by H. Neave, with permission from Unwin Hyman Limited.

the tables. Basically the tables work similarly to the process shown in Figure 4. The 95% (or P 0.05) column represents the expected 1:20 chance of there being no difference (equal to line A in Figure 4), while the 99% (or P 0.01) gives us the 1:100 theoretical value of no difference at this high level of significance (equivalent to line B in Figure 4). So in a similar way, t-test values to the left of the critical figure (i.e., lower numerically) indicate no significant difference, while figures to the right of the critical value mean we reject the null hypothesis and accept that there is a significant difference. For example, looking at Table 1 for 5 degrees of freedom, a calculated t-test value of 3.36 would be significantly different at the 5% level of confidence (P 0.05 = 2.57) but would not be significantly different at the 1% level (P 0.01 = 4.03). The difference in the two values obtained from the calculated t value suggest that there is a significant but not highly significant difference in the data.

STATISTICAL EVALUATION OF DATA

Before we begin our test it is necessary to decide whether to use a one-tailed or two-tailed test, sometimes called one-sided or two-sided tests. This can be easily done by answering the question Do we want to know if the two means are different from each other (a two-tailed test), or do we want to know if one mean is either greater than or less than another (a one-tailed test)? Most t tables give the critical values for a range of confidence levels for two-tailed tests. For a one-tailed test the value in the table should be taken for the column for twice the desired confidence level; for example, for a one-tailed test, $P = 0.05$, 5 degrees of freedom, the critical value is read from the $P = 0.10$ column, and in the example used above this would correspond to 2.02.

Only brief mention has been made of the F test, but it is appropriate to bring it in at this point. The F test has a very important role in our significance testing, confirming whether or not the data distribution (or variance) for methods being compared for accuracy (t test) are different. It may be necessary to establish this fact before attempting to determine the accuracy of the t test.

F Test for the Comparison of Precisions

The F test simply expresses the two sample variances (the standard deviation squared) as a ratio:

$$F = S_1^2 / S_2^2 \tag{6}$$

where S_1^2 is the variance of data set 1 and S_2^2 is the variance of the data set 2.

The denominator and numerator in Eq. (6) may have to be reversed in order to give an F value ≥ 1. If the null hypothesis is true and there is no significant difference in the two variances, the ratio should be close to being 1. The critical or predicted values for various levels of significance can be found in Table 2. The requirement for selecting a one-tailed or two-tailed test also applies to this test.

Example 1. The data for the percentage yield of an analyte obtained by two methods are given below.

Method A (%)	Method B (%)
79.4	78.0
77.1	81.2
76.2	80.5
77.5	78.2
78.6	79.8
77.7	79.5

Table 2 Critical Values of F for a One-Tailed Test ($P = 0.05$)

v_1	1	2	3	4	5	6	7	8	9	10	12	15	20
v_2													
1	161.4	199.5	215.7	224.6	230.2	234.0	236.8	238.9	240.5	241.9	243.9	245.9	248.0
2	18.51	19.00	19.16	19.25	19.30	19.33	19.35	19.37	19.38	19.40	19.41	19.43	19.45
3	10.13	9.552	9.277	9.117	9.013	8.941	8.887	8.845	8.812	8.786	8.745	8.703	8.660
4	7.709	6.944	6.591	6.388	6.256	6.163	6.094	6.041	5.999	5.964	5.912	5.858	5.803
5	6.608	5.786	5.409	5.192	5.050	4.950	4.876	4.818	4.772	4.735	4.678	4.619	4.558
6	5.987	5.143	4.757	4.534	4.387	4.284	4.207	4.147	4.099	4.060	4.000	3.938	3.874
7	5.591	4.737	4.347	4.120	3.972	3.866	3.787	3.726	3.677	3.637	3.575	3.511	3.445
8	5.318	4.459	4.066	3.838	3.687	3.581	3.500	3.438	3.388	3.347	3.284	3.218	3.150
9	5.117	4.256	3.863	3.633	3.482	3.374	3.293	3.230	3.179	3.137	3.073	3.006	2.936
10	4.965	4.103	3.708	3.478	3.326	3.217	3.135	3.072	3.020	2.978	2.913	2.845	2.774
11	4.844	3.982	3.587	3.357	3.204	3.095	3.012	2.948	2.896	2.854	2.788	2.719	2.646
12	4.747	3.885	3.490	3.259	3.106	2.996	2.913	2.849	2.796	2.753	2.687	2.617	2.544
13	4.667	3.806	3.411	3.179	3.025	2.915	2.832	2.767	2.714	2.671	2.604	2.533	2.459
14	4.600	3.739	3.344	3.112	2.958	2.848	2.764	2.699	2.646	2.602	2.534	2.463	2.388
15	4.543	3.682	3.287	3.056	2.901	2.790	2.707	2.641	2.588	2.544	2.475	2.403	2.328
16	4.494	3.634	3.239	3.007	2.852	2.741	2.657	2.591	2.538	2.494	2.425	2.352	2.276
17	4.451	3.592	3.197	2.965	2.810	2.699	3.614	2.548	2.494	2.450	2.381	2.308	2.230
18	4.414	3.555	3.160	2.928	2.773	2.661	2.577	2.510	2.456	2.412	2.342	2.269	2.191
19	4.381	3.522	3.127	2.895	2.740	2.628	2.544	2.477	2.423	2.378	2.308	2.234	2.155
20	4.351	3.493	3.098	2.866	2.711	2.599	2.514	2.447	2.393	2.348	2.278	2.203	2.124

Critical Values of F for a Two-Tailed Test ($P = 0.05$)

v_1	1	2	3	4	5	6	7	8	9	10	12	15	20
v_2													
1	647.8	799.5	864.2	899.6	921.8	937.1	948.2	956.7	963.3	968.6	976.7	984.9	993.1
2	38.51	39.00	39.17	39.25	39.30	39.33	39.36	39.37	39.39	39.40	39.41	39.43	39.45
3	17.44	16.04	15.44	15.10	14.88	14.73	14.62	14.54	14.47	14.42	14.34	14.25	14.17
4	12.22	10.65	9.979	9.605	9.364	9.197	9.074	8.980	8.905	8.844	8.751	8.657	8.560
5	10.01	8.434	7.764	7.388	7.146	6.978	6.853	6.757	6.681	6.619	6.525	6.428	6.329
6	8.813	7.260	6.599	6.227	5.988	5.820	5.695	5.600	5.523	5.461	5.366	5.269	5.168
7	8.073	6.542	5.890	5.523	5.285	5.119	4.995	4.899	4.823	4.761	4.666	4.568	4.467
8	7.571	6.059	5.416	5.053	4.817	4.653	4.529	4.433	4.357	4.295	4.200	4.101	3.999
9	7.209	5.715	5.078	4.718	4.484	4.320	4.197	4.102	4.026	3.964	3.868	3.769	3.667
10	6.937	5.456	4.826	4.468	4.236	4.072	3.950	3.855	3.779	3.717	3.621	3.522	3.419
11	6.724	5.256	4.630	4.275	4.044	3.881	3.759	3.664	3.588	3.526	3.430	3.330	3.226
12	6.554	5.096	4.474	4.121	3.891	3.728	3.607	3.512	3.436	3.374	3.277	3.177	3.073
13	6.414	4.965	4.347	3.996	3.767	3.604	3.483	3.388	3.312	3.250	3.153	3.053	2.948
14	6.298	4.857	4.242	3.892	3.663	3.501	3.380	3.285	2.209	3.147	3.050	2.949	2.844
15	6.200	4.765	4.153	3.804	3.576	3.415	3.293	3.199	3.123	3.060	2.963	2.862	2.756
16	6.115	4.687	4.077	3.729	3.502	3.341	3.219	3.125	3.049	2.986	2.889	2.788	2.681
17	6.042	4.619	4.011	3.665	3.438	3.277	3.156	3.061	2.985	2.922	2.825	2.723	2.616
18	5.978	4.560	3.954	3.608	3.382	3.221	3.100	3.005	2.929	2.866	2.769	2.667	2.559
19	5.922	4.508	3.903	3.559	3.333	3.172	3.051	2.956	2.880	2.817	2.720	2.617	2.509
20	5.871	4.461	3.859	3.515	3.289	3.128	3.007	2.913	2.837	2.774	2.676	2.573	2.464

v_1 = number of degrees of freedom of the numerator and v_1 = number of degrees of freedom of the denominator.

Tabular extracts taken from *Elementary Statistics Tables* by H. Neave, with permission from Unwin Hyman Limited.

STATISTICAL EVALUATION OF DATA

Method A	Method B
$\bar{x} = 77.7$	$\bar{x}_2 = 79.5$
$S_1 = 1.12$	$S_2 = 1.26$
$S_1^2 = 1.25$	$S_1^2 = 1.58$
$n = 6$	$n = 6$

Is the precision of the two methods significantly different?
Calculate F so the value is ≥ 1:

$$F = S_2^2/S_1^2 = 1.58/1.25 = 1.26$$

Because we want to know if the variance of one method is greater or less than the second, we select a two-tailed test. The critical value is obtained from Table 2. The degrees of freedom are 5 in both cases, and the conversion is that we read the degrees of freedom for the numerator (in this case S_2^2) across the top of the table and the degrees of freedom for the denominator (in this case S_1^2) down the side of the table.

$F_{5,5}$ critical value for two-tailed test (degrees of freedom 5,5) at the 5% level of significance = 7.146

As our test value is less than the critical value (i.e., 1.26 test, 7.146 crit), we conclude that there is no significant difference in the precisions of the two methods at the 5% level of confidence.

Example 2. Let us suppose we had a reference method (method A) and a faster modified method (method B) and wanted to know if the precision of the modified method (B) were significantly better than that of the reference method A (a one-tailed test).

Method A	Method B
$\bar{x}_1 = 52$	$\bar{x}_2 = 51$
$S_1 = 3.41$	$S_2 = 1.72$
$S_1^2 = 11.62$	$S_2^2 = 2.95$
$n = 9$	$n = 10$

$$F = S_1^2/S_2^2 = 11.62/2.95 = 3.94$$

Note that in this case the degrees of freedom are different due to one more value being taken for method B.

$F_{8,9}$ crit value for one-tailed test (degrees of freedom 8,9) at the 5% level of significance = 3.230

Since the test value (3.94) exceeds the critical value (3.230), the variance of the modified method B is significantly smaller than that of the proposed method at the 5% level of confidence; that is, the modified method is more precise.

Examples 1 and 2 show how you can use the F test to evaluate the precision of a method and to confirm that there is no significant difference in the variance of two methods. This latter point is important to establish if we wish to proceed with a t test for accuracy, against a reference method.

Notes of caution with the F test:

1. Ensure that your ratio of the variances gives a value ≥ 1.
2. When looking up the degrees of freedom in the tables, make sure you have them the same way around as your numerator and denominator values in the ratio calculation.
3. Be sure to use the correct critical value when using a one- or two-sided test.

t Test for the Estimation of Accuracy

The analytical chemist uses the t test in three basic ways.

Comparison of a Mean with a Certified Value

Perhaps the simplest use of the t test is when we examine our experimental data $\bar{x}$ for significant differences against a referee value μ.

Example 3. We have the following data:

$\bar{x} = 85$ (obtained from $n = 10$ replicates)
$S = 0.6$
$\mu = 83$ (reference value)

$t = (\bar{x} - \mu) \sqrt{n} / S$
$t = 85 - 83 \sqrt{10}/0.6 = 10.5$ \hfill (7)

From Table 1 the t critical value for a one-sided test (Is the 85 significantly *higher* than 83?) at the 5% level for $n-1$ degrees of freedom will be 1.83. Since the test value is greater than the critical value, we conclude that the experimental datum is significantly higher than the reference value but that the method did have good precision.

Comparison of the Means from Two Samples, One of Which May Be a Reference Method

Example 4. Let us suppose we have mean data for a reference method A ($\bar{x}_1$) and a modified method B ($\bar{x}_2$), and we want to know if the modified method is statistically different from the reference method. The null hypothesis assumes that there is no significant difference between the two means; that is, $\bar{x}_1 - \bar{x}_2$ should not differ significantly from zero. We recall from the previous section that we should establish that the variances of the two methods are not significantly different before we proceed with our t test; this requires us to do an F test on the data first.

Reference method A	Modified method B
$\bar{x}_1$ 6.40	$\bar{x}_2$ 6.56
S_1 0.126	S_2 0.179
S_1^2 0.015	S_2^2 0.032
$n = 10$	$n = 10$

$$F = S_1^2/S_2^2 = 0.032/0.015 = 2.13$$

Two-tailed test $F_{9,9}$ crit, 5% level of significance $= 4.026$

There is no significant difference in the precisions of the two methods at the 5% level.

We can now proceed with the t test. As the two samples have standard deviations that do not significantly vary, a pooled estimate of their standard deviations can be calculated from the two individual standard deviations using the equation

$$S^2 = \frac{(n_1 - 1)S_1^2 + (n_2 - 1)S_2^2}{n_1 + n_2 - 2} \tag{8}$$

Thus

$$S^2 = \frac{9 \times 0.015 + 9 \times 0.032}{18} = 0.0235$$

$$S = 0.153$$

Our t value is determined from

$$t = \frac{\bar{x}_1 - \bar{x}_2}{S(1/n_1 + 1/n_2)^{1/2}} \tag{9}$$

thus

$$t = \frac{6.40 - 6.56}{0.153 \, (1/10 + 1/10)^{1/2}} = -2.35$$

There are 18 degrees of freedom, so the critical value for t at the 5% level of significance (two-tailed test) is 2.10.

Note that the t value is $\pm$ so the negative value in this case carries no importance.

Since the test value 2.35 is greater than the critical value 2.10, we can say that there is a significant difference between the two results at the 5% level of confidence; that is, the modified method, while being as precise as the reference method, is not as accurate. This does assume, however, that the reference method is accurate, a point discussed earlier.

The Paired t Test

When developing or evaluating a method it is common to analyze a range of samples that will have varying concentrations for a particular analyte. We therefore end up with a data set for, say, 10 different samples that may have been prepared by two methods. Comparing two means as in the previous section would not be appropriate, as the variation in data (due to the different samples being analyzed) would be too great. However, we could test to see if the differences between the pairs of values for each sample obtained by the two methods were significantly different. An F test is not appropriate in this case, as clearly the values from different samples cannot be considered to have a variance from the same population.

Example 5. Consider the data on 10 samples gathered by two methods given in Table 3.
Are the data for the two methods significantly different?
Since μ_d should equal 0, the t value is obtained from

$$\begin{aligned} t &= \bar{x}_d \sqrt{n}/S_d \\ t &= 0.4\sqrt{10}/4.029 = 1.962 \end{aligned} \tag{10}$$

The critical value for t at the 5% level of significance with $n-1$ degrees of freedom from the table is 2.26, and the test is two-tailed. Since the test value is less than the critical value, we can conclude that methods 1 and 2 show no significant difference.

In the paired t test we have made the assumption that any errors, either random or systematic, are independent of the concentration. Over very wide

STATISTICAL EVALUATION OF DATA

Table 3 Method for Calculating a Paired t test on Data from Two Different Methods of Analysis for Ten Separate Samples

Sample	Method 1	Method 2	Difference d	$d - \bar{X}_d$	$(d - \bar{X}_d)^2$
1	90	87	3	2.6	6.76
2	30	34	−4	−4.4	19.36
3	62	60	2	1.6	2.56
4	47	50	−3	−3.4	11.56
5	61	63	−2	−2.4	5.76
6	53	48	5	4.6	21.16
7	40	38	2	1.6	2.56
8	88	80	8	7.6	57.76
9	76	78	−2	−2.4	5.76
10	10	15	−5	−5.4	29.16
			$\Sigma = 4$		$\Sigma = 162.4$
			$\bar{X}_d = 0.4$		$S_d = \sqrt{162.4/10}$
					$= 4.029$

concentration ranges this assumption may not hold true, and techniques such as linear regression should be applied to the data; see Chapter 5 for details.

In conclusion, we have seen three ways in which the t test can be used to give us a statistical evaluation of our data. It is important to remember that when we perform the test all we are really estimating is whether there is or is not a difference at some selected level of significance between two means. It is important, therefore, that we evaluate the results from this type of significance testing in an intelligent way and not distort the statistical inferences of our results. There exists the possibility that we will reject a test value at the 5% level of confidence even if it is true. Such an outcome is known as type I error. The changes of making such an error can be reduced by moving our test level of significance to, say, the 1% or even 0.1% level of confidence. In a similar way it is also possible that we could accept a value as not being significantly different when in fact it is; this is known as a type II error. To get a graphical impression of these two errors, if we look again at Figure 4 and the line of significance, then to the right of the line we have the possibility of type I errors and to the left of the line type II errors. While the two types of errors are obviously different, increasing the significance level to reduce the risk of type I error will increase the risk of type II error, and vice versa. The obvious way to minimize the possible problem of these potential errors is to sharpen up the distribution profiles or, if you like, get better resolution of the μ and x distributions. This can be achieved by increasing the number of samples taken, which will decrease the standard error of the mean. The analyst should not be overly worried by the statistical arguments regarding these errors, but they are mentioned in conclusion to illustrate the limitations of significance testing.

The examples given have all been for the case when two mean values are known. In our analytical work, however, we often end up with more than two means to be compared. We therefore require a slightly different approach to significance testing and adopt a technique known as the analysis of variance, the subject of the next section.

2.4 ANALYSIS OF VARIANCE

We have seen from the previous section how a sample t test can be used to estimate the statistical significance of mean results obtained for, say, two methods. We may wish, however, to compare the results for more than two means—for example, results from four different laboratories or from three different methods carried out in the same laboratory. With such data there are two possible sources of variation, those associated with systematic errors or the fixed-effect factor and those arising from random errors as described previously. *Analysis of variance* (or ANOVA for short) is a technique that will allow us to evaluate such data for the influence of several poosible effects operating simultaneously [6]. In practice we use ANOVA in two main ways. The first is to separate any variation in our data that is caused by changing the fixed-effect factor or controlled factor from the variation due to random error. In this instance the term *factor* is defined as any variable likely to affect the outcome of an experiment. This approach will enable us to estimate the infuence that changing the controlled factor (e.g., the method of analysis used) will have on our errors. The second practical use of ANOVA is to identify more than one source of random error, for example, sampling and inhomogeneity; this variation is sometimes known as a *random-effect factor*. Using ANOVA in these two ways we are primarily concerned with one factor either controlled or random and a random error in the measurement; these types of analysis are known as *one-way* ANOVA. This terminology is used to distinguish the tests described here from two-way ANOVA, a technique used primarily in experimental design (see Chapter 6).

Let us look now more closely at how we use one-way ANOVA to evaluate our data.

ANOVA to Test for Differences Between Means

Consider the evaluation of four different extraction procedures (the fixed-effect factor) for an organic compound in water determined by its UV absorbance. The analytical chemist will have prepared a test solution of the organic substance in water and will perform replicate extractions with each of the proposed proce-

STATISTICAL EVALUATION OF DATA

dures and obtain an absorbance value. The following shows the kind of results that could be obtained from such a set of experiments.

Extraction method	Replicate measurements (arbitrary units)	Mean
A	300, 294, 304	299
B	299, 291, 300	296
C	280, 281, 389	283
D	305, 310, 300	305
	Overall mean	296

We see that the mean values for the four extraction methods are different. However, we are aware that random error may cause one sample mean to vary from the next. We use ANOVA to test whether the differences between the extraction methods are too great to be explained by the random error. To do this we return to our null hypothesis, which assumes that our extraction methods are drawn from a population with mean μ and variance σ^2. On this basis the variance can be estimated in two ways—within-sample variation and between-sample variation,. This is where we start our ANOVA.

Within-Sample Variation

For each extraction method the variance V can be calculated using an equation similar to that with which we calculate standard deviation, but in this case we do not take the square root:

$$V = \frac{\Sigma(x_1 - \bar{x})^2}{n - 1} \tag{11}$$

From our data we get the following:

$$\text{Variance of method A} = \frac{(300-299)^2 + (294-299)^2 + (304-299)^2}{3 - 1} = 25.5$$

$$\text{Variance of method B} = \frac{(299-296)^2 + (291-296)^2 + (300-296)^2}{3 - 1} = 25.0$$

$$\text{Variance of method C} = \frac{(280-283)^2 + (281-283)^2 + (289-283)^2}{3 - 1} = 24.5$$

$$\text{Variance of method D} = \frac{(305-305)^2 + (310-305)^2 + (300-305)^2}{3 - 1} = 25.0$$

Averaging these values gives the within-sample estimate of the variance

$(25.5 + 25.0 + 24.5 + 25.0)/4 = 25$

This is known as the *mean square* because it is an average of the squared terms.

As we see, each of the mean-squared terms has one degree of freedom. When we sum these means we have to take into account an additional degree of freedom for each term. This means that for the overall sum of the squared terms we have $12 - 4 = 8$ degrees of freedom and obtain a value of

$25 \times 8 = 200$

Between-Sample Variation

We calculate the between-sample variation in a similar way to the within-sample variation; see Eq. (11).

Method mean variance =
$$\frac{(299-296)^2 + (296-296)^2 + (283-296)^2 + (305-296)^2}{4 - 1} = 86$$

Thus the between-sample estimate of the variance, correcting for degrees of freedom, will be

$86/3 \times 3 = 86$

So we have now calculated the following:

Within-sample mean square = 25 with 8 d.f.

Between-sample mean square = 86 with 3 d.f.

We now use a one-tailed F test to see if the between-sample estimate of the variance is greater than the within-sample estimate of the variance. If there is no significant difference, then these two variances should not differ significantly. If, on the other hand, the between-sample variation is greater than the critical value, it indicates a significant difference in the means.

Applying the F test we get

$F = 86/36 = 3.44$

(Note that we do not square the values because we are using the mean square.)

From Table 2 we get a critical value of $F_{3,8} = 4.066$ ($P = 0.05$). Since the test value of F is less than the critical value, the null hypothesis is retained, and we conclude that the method means do not differ significantly.

A significant result in our one-way ANOVA would be indicative of various problems. For example, anything from one mean being different to all the means differing could have occurred. A simple way of estimating the difference between adjacent values is to calculate the least significant difference.

STATISTICAL EVALUATION OF DATA

$$S \sqrt{2/n} \times t_{h(n-1)} \qquad (12)$$

where S is the within-sample variance, h is the number of samples, $h(n-1)$ is the total number of degrees of freedom, and t is the critical value from the table for the total number of degrees of freedom.

Let us look at a brief example.

Say, in a further experiment, instead of extraction methods, storage times for our reagent were examined which might have been, freshly prepared (A), one day old (B), one week old (C), one month old (D). The following data provide an example of the types of results that could typically be obtained.

Storage of reagent	$\bar{x}$
A fresh	51
B 1 day	52
C 1 week	49
D 1 month	42

Let us suppose that one-way ANOVA for these results indicated that the means of A–D differ significantly. We now wish to identify the nature of this variation between the means.

First, arrange the means in increasing order of size:

$$\bar{x}_D = 42, \qquad \bar{x}_C = 49, \qquad \bar{x}_A = 51, \qquad \bar{x}_B = 52$$

Calculating the least significant difference, from Eq. (12),

$$\sqrt{3} \times \sqrt{2/3} \times 2.31 = 3.26$$

where the within-sample mean square = 3 with 8 d.f. and t_8 crit (P 0.05) = 2.31

We can now see numerically whether the differences between pairs of means are greater than our least significant difference of 3.26. Looking in more detail at the pairs, we get

$\bar{x}$ for storage test	Difference
D/C	5
D/A	9
D/B	10
C/A	2
C/B	3
A/B	1

We see from this table that D differs significantly from A, B, and C and that C is very close to the test value. This suggests that the time of storage is important and that after 1 week the reagent apparently "goes off."

ANOVA along with many statistical tests shows a repetitive nature in the calculations that enables us to practically apply a shorter, simpler form to the calculation. In this case this involves summing the squares of the deviations from the overall mean and dividing by the number of degrees of freedom. Assessing the total variance in this way takes into account both the within- and between-samples variations, but fortunately there is a clearly defined algebraic relationship relating the two variances. This relationship basically states that the sum of the between- and within-sample variations will equal the total, so by calculating the between-sample variation and the total variation we can obtain the within-sample variation by subtraction.

The general approach to carry out the method of calculation can be summarized as follows.

Source of variation	Sum of squares	Degrees of freedom
Between-samples	$\Sigma T_i^2/n - T^2/N$	$h-1$
Within-samples	By subtraction	By subtraction
Total	$\Sigma x^2 - T^2/N$	$N-1$

where N is the total number of measurements, n the number of replicate measurements for each sample, h the number of samples, T_i the sum of the measurements for the ith sample, T the grand total of all measurements, and Σx^2 the sum of the squares of all the data points.

To illustrate how this approach to calculating ANOVA can be carried out, an example from the second problem we discussed earlier will be used. In this example, one-way ANOVA will be used to test for differences where a random effect, rather than a fixed effect, is suspected that is a sampling, not a systematic, variation. In this case we are using ANOVA to identify and estimate the different random sources of variation that occur with both sampling and analysis. An example may be the determination (four replicates) of arsenic in coal for samples A–E collected from different parts of a ships hold. In this case we would generate data as follows.

Sample	Arsenic content (ng/g)	Mean
A	72, 73, 72, 71	72
B	73, 74, 75, 73	74
C	74, 75, 74, 76	75
D	71, 72, 71, 73	72
E	76, 75, 71, 76	75

The first step is to calculate the mean squares using the general formula given above. To make the arithmetic simpler, 70 has been subtracted from all the arsenic values.

Sample	Data	T	T^2
A	2 3 2 1	9	81
B	3 4 5 3	15	225
C	4 5 4 6	15	225
D	1 2 1 3	7	49
E	6 5 7 6	24	576
		$\Sigma T = 70$ $\Sigma T^2 = 1156$	

In this case, $n = 4$, $h = 5$, $N = 20$, and $\Sigma x^2 = 331$.

Substituting these values into our ANOVA table we get

	Sum of squares	d.f.	Mean squares
Between-sample	$1156/4 - 70^2/20 = 44$	4	$44/4 = 11$
Within-sample	42	15	$42/15 = 2.8$
Total	$331 - 70^2/20 = 86$	19	

We see that the between-sample mean square is greater than the within-sample mean square. Does this mean that the sampling variance is significantly different from the variance associated with arsenic determination? To test, we apply the F test and compare the two mean squares.

$F = 11/2.8 = 3.928$

From table 2 we see that the critical $F_{4,15}$ value at the 5% level for a two-tailed test is 3.804. One can conclude, therefore, that there is a significant difference in the random sampling error over the random analytical error. In passing, it is interesting to note that we have observed in this example that the precisions of the sampling and analysis are significantly different. Can we therefore use this type of test to assist us in our sampling strategy?

The first thing to do may be to examine more closely the actual estimates of precision or the between- and within-sample variances. From the calculation above we already have an estimate of the within-sample variance, 2.8. However, the between-sample mean square estimates the within-sample plus the between-sample variance, so to obtain an estimate of the between-sample variance only we perform the simple subtraction

Estimate of
between-sample variance = (between-sample mean square
within-sample mean square)/n
= (11 − 2.8)/4
= 2.05

So we summarize:

Estimate of between-sample variance = 2.8
Estimate of within-sample variance = 2.05

These estimates of the relative precisions indicate that they are not greatly different in magnitude, and this is also shown by how close the F test and F critical values came out. In other words, making some value judgment on the results, the analyst may conclude that the between-sample variance was only just worse than the within-sample variance at the 5% level of significance, suggesting that the samples of coal taken from the ship's hold were reasonably similar. The better the precision of our technique, then, the smaller will be our variance, and, as we expect, the analytical determination (replicate determinations on a sample) should in most cases be better than the sampling variance. However, there is little point in attempting to improve the analytical variance to much more than one-tenth of the sampling variance, since any further reduction will not greatly improve the total variance (which is the sum of the two variances). As a final point, the number of replicate samples (n) taken for the analysis will decrease the confidence interval, and it is common in bulk-sampling techniques to bulk samples for blending before making n replicate measurements.

The confidence interval is often used to estimate the range within which a true mean may be found. The limits for the confidence interval are known as the confidence limits and are given by the expression

$$\text{Confidence limits of } \mu = \bar{x} \pm t(S/\sqrt{n}) \tag{13}$$

where t is a value from the t table for a certain level of significance and degrees of freedom.

In conclusion, we have seen how ANOVA can be a very powerful technique for evaluating both systematic and random errors in our data. Furthermore, it can also identify potential sources of error in such data and can therefore assist the analyst in ensuring that he or she is using good reliable analytical methodology.

2.5 OUTLIERS

The significance tests we have been using depend very much on the distribution of the variance in our data. One of the problems that frequently arise is the

STATISTICAL EVALUATION OF DATA

Table 4 Critical Values of Q ($P = 0.05$)

Sample size	Critical value
4	0.831
5	0.717
6	0.621
7	0.570
8	0.524
9	0.492
10	0.464

Tabular Extracts taken from *Elementary Statistics Tables* by H. Neave, with permission from Unwin Hyman Limited.

rejection of outliers or the erroneous result. If we leave them in, they may distort an otherwise good data set; however, if we remove them, are we biasing the outcome of our tests? Deciding on what is a true outlier is obviously complicated and will often depend on the nature of the distribution. For the purpose of this discussion, only normally distributed data will be considered. It may be of some interest, however, to look at the next chapter, where it is shown that nonparametric methods can be used very effectively to identify outlier or rogue data points.

A popular and well-tested way of evaluating a measurement as an outlier is to use Dixon's Q test. This test compares the difference between a suspect measurement and the measurement nearest to it in size with the difference between the highest and lowest values, to produce a ratio of these differences.

$$Q = \frac{\text{suspect value} - \text{nearest value}}{\text{largest value} - \text{smallest value}}$$

Table 4 gives the critical values for Q at the 5% level of significance. As with the t test, if the test Q value exceeds the critical value, then the suspect data point should be rejected. For example, take the data set

11.2, 12.1, 16.9, 12.4, 13.6, 12.6, 13.2

The third data point is suspect. Should it be rejected?

$$Q = \frac{16.9 - 13.6}{16.9 - 11.2} = \frac{3.3}{5.7} = 0.579$$

From Table 4, for sample size 7 the critical Q value at the 5% level of significance is 0.570. Since the test value exceeds the critical value, the 16.9 data point should be rejected from the data set. If, after removal of an outlier, the data still appear to have a rogue value, then the test can be repeated. For example, we may wonder if the value of 13.6 is an outlier.

$$Q = \frac{13.6 - 13.2}{13.6 - 11.2} = 0.16$$

The test value of 0.16 is now lower than the critical value of 0.621 (Q 5% significance for sample size 6), and so we conclude that 13.2 is not an outlier. It is interesting to note that our value of 16.9 is only just outside the significant test value.

Remember that at the 5% level of significance there remains a 1 in 10 chance of incorrectly rejecting a suspect value. For the above example, the mean standard deviation and relative standard deviation with and without the outlier would be

	$\bar{x}$	S	RSD
With outlier	13.1	1.8	13.9%
Without outlier	12.5	0.84	6.7%

There are various other techniques available to distinguish outliers, but the practices can be fraught with danger. The simple example given above serves to illustrate the usual situation of a misrecorded value, bad sample injection, or instrumental spike. The Q test is helpful in modifying our data for further analysis, but the analyst must be conscious that if a value is removed from the data its removal must be reported in the final evaluation of the results. Details of the Q test and other discriminative tests can be found in reference 7.

2.6 TESTING FOR NORMALITY OF DISTRIBUTION

The techniques described in this chapter are all based on the assumption that the data are normally distributed. As briefly indicated there are numerous other forms or distributions, and often, given the small number of data points, the analyst's certainty of the nature of the distribution cannot be confirmed.

One method where a large data set is available (50 or more data points) is the chi-squared test. We may use such a test, for example, to estimate if the down times on instruments from three different manufacturers differ over the year.

Say the numbers of days down time for the three instruments are 20, 14, and 8. If we adopt the null hypothesis, then we will assume that there is no difference in the reliability of the three instruments. Since the total number of days down is 42, the expected average number of down days for any one instrument will be 42/3 = 14. To give us units in whole days we adjust the distribution

STATISTICAL EVALUATION OF DATA

Table 5 Critical Values of χ^2 ($P = 0.05$)

Number of degrees of freedom	Critical value
1	3.84
2	5.99
3	7.81
4	9.49
5	11.07
6	12.59
7	14.07
8	15.51
9	16.92
10	18.31

Tabular extracts taken from *Elementary Statistics Tables* by H. Neave, with permission from Unwin Hyman Limited.

to be 14, 14, 14. The question is, Are the differences between the observed instrumental days down time and the expected frequencies different enough to be rejected by the null hypothesis? In the chi-squared (χ^2) test, the following assessment is made.

Observed frequency, O	Expected frequency, E	O−E	(O−E)²/E
20	14	6	2.57
14	14	0	0
8	14	−6	2.57
		0.00	$\chi^2 = 5.14$

Note that the total of the O−E column should always equal zero. If χ^2 exceeds the critical value, then the null hypothesis will be rejected. As with other tests, we have to allow for the lack of independence in the data by using a number of degrees of freedom one less than the number of classes—in this case 3−1, or 2. The critical value χ^2 for the 5% level of significance and 2 degrees of freedom can be found in Table 5. The critical value of 5.99 is larger (only just) than the test value of 5.14, which indicates that the null hypothesis is accepted; that is, there is no evidence that the down times on the three instruments differ significantly at the 5% level of confidence. Again the analyst must be careful not to draw more conclusions from such a result than are reasonable; I would choose the instrument with a down time of only 8 days over one with 20 days. In the case above, the *t* test would not be appropriate because we are dealing with frequency—the number of times an event happens—rather than continuous variates in a distribution. In general, the chi-squared test should only be applied to a data set in which we are attempting to test a frequency of some event and where there are 50 or more observations.

We see from the above example that the chi-squared test can be used to indicate if observations differ significantly from a normal distribution. However, the method often requires a data set larger than the experimental analyst has available. A relatively simple way to assess whether or not a set of data follows a normal distribution is to plot a cumulative frequency curve on special graph paper known as normal probability paper, which converts the typical S-shaped curve into a straight line. The straightness of the line can then be assessed using least squares techniques to give a fit index of normality for the data. This is commonly done using the correlation coefficient r. The assumption that our data follow a normal or Gaussian distribution is fundamental in our use of the significance tests described in this chapter. If there is any reason to suspect that your data follow an alternative distribution, then it is important to examine the data; it is often the case that non-normal distributions can be transposed into normal distributions or that alternative tests can be used [8, 9].

2.7 COMPARISON OF TWO METHODS BY LEAST SQUARES FITTING

As we have seen, it is often the case that we have a set of experimental data obtained for, say, both a new and a reference technique. In the absence of error, the plot of one set of data against the second should have a slope of exactly unity and an intercept of zero. Because we are able to numerically estimate variations in the best fitting of a line by a least squares technique we have a useful method to assess our data. For details of the least squares technique, see Chapter 4. To illustrate how this graphical method can be used, we see in Figure 4 the kinds of errors that are possible. The random error can be estimated from the calculation of the standard deviation in the y direction S_y/x (sometimes called the standard deviation of the estimate of y on x). As we see from Figure 4, we are able to identify three types of error—random, proportional, and constant—by this least squares approach using a goodness-of-fit test. In reality, when we regress our modified method data (y) on our reference data (x) they are both subject to random errors, and we strictly should use the orthogonal regression technique. But for practical reasons we test the two methods against each other, y as a function of x and x as a function of y, assuming that x and y are free from random error in the respective cases. In this way we can employ least squares regression. To test whether the slope of the best fit line (m) is significantly different from unity, a t test can be used.

$$t = \left(\frac{m-1}{\sqrt{1-r^2}}\right) \times \sqrt{n-2} \qquad (14)$$

where m is the actual slope and r is the correlation coefficient. The test will have $n-2$ degrees of freedom.

In addition, a simple test can be used to assess whether or not the intercept (C) significantly differs from O.

$$t = (C-O)/S_n^2 \qquad (15)$$

where C is the intercept, S_n is the standard deviation of replicate intercept measurement, and there are $n-2$ degrees of freedom.

The t-test values are looked up in the usual way, and evaluations are made.

The least squares approach is a simple alternative method of evaluating data that the reader may wish to consider. Further information can be found in references 10 and 11.

ADDITIONAL RECOMMENDED READING

Miller, J. C., and J. N. Miller, *Statistics for Analytical Chemistry*, 2nd ed., Ellis Horwood Series in Analytical Chemistry, Wiley, New York, 1988.
Huff, D., *How To Be with Statistics*, Penguin, New York, 1954.
Barlow, F. J., *Statistics*, Wiley, New York, 1989.
Chatfield, C., *Statistics for Technology*, Chapman and Hall, London, 1983.
Moroney, M. J., *Facts from Figures*, 3rd ed., Penguin, New York, 1952.

REFERENCES

1. B. E. Cooper, *Statistics for Experimentalists*, Pergamon, London, 1969.
2. R. Smith and G. V. Jones, *The Sampling of Bulk Materials*, Published by the Royal Society of Chemistry, London, 1981.
3. IUPAC, *Compendium of Analytical Nomenclature*, Pergamon,
4. Guide for use of terms in reporting data, *Anal. Chem.*, 54: 157 (1982).
5. W. T. Eadie et al. *Statistical Methods in Experimental Physics*, Elsevier/North-Holland, Amsterdam, 1971.
6. G. E. P. Box et al., *Statistics for Experimentalists*, Wiley, New York, 1978.
7. V. Barnett and T. Lewis, *Outliers in Statistical Data*, Wiley, New York, 1984.
8. B. Kowalski, *Chemometrics—Theory and Application*, American Chemical Society Chapter 11, Washington, D.C., 1977.
9. D. L. Massart, B. G. M. Vandeginste, S. D. Deming, Y. Michatte, and D. C. Kaufman, *Chemometrics: A Textbook*, Chapter 3, Elsevier, New York, 1987.
10. G. W. Snedecor and W. G. Cochran, *Statistical Methods*, Iowa State Univ., Ames, 1989.
11. N. R. Draper and H. Smith, *Applied Regression Analysis*, Wiley, New York, 1981.

3
Exploratory, Robust, and Nonparametric Data Analysis

John Michael Thompson *University of Keele Hospital Centre, Stoke on Trent, England*

Classically, statistical techniques have been taught to science and technology students on the basis of strong assumptions about data behavior. For measurements of continuous variables (e.g., absorbance, weight), the assumption would have been that the data were distributed in a Gaussian (normal) manner or according to some other well-defined distribution, as illustrated in Figure 1. For discrete data (e.g., photon or particle counts; categories or characteristics such as red/blue or male/female), appropriate distributions might be Poisson or binomial, for example (see Figure 2). Statistical techniques that make strict assumptions about the distribution that the experimental data are expected to follow are termed *parametric* methods.

Conover [1] describes parametric statistics as "the means of obtaining exact solutions to approximate problems." Statisticians have devised models (e.g., the Gaussian or normal distribution, shown in Figure 1) for "populations" from which samples (i.e., collections) of data are drawn, so that exact probability functions can be formulated. Data from real experiments, which in general conform only approximately to such models, are then analyzed to determine statistical parameters of the data (such as mean, variance, and standard deviation) as if they fitted the model. Well-known tests, such as Student's t-test and the Fisher F-test, are applied on the assumption that the data fit a Gaussian distribution or can be simply transformed to one (e.g., by taking logarithms or square roots).

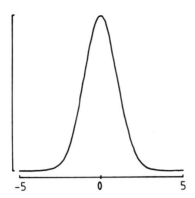

Figure 1 Plot of the frequency of measurements against their magnitude for a Gaussian distribution.

In the late 1930s, a radically different approach was developed that involved few, if any, changes to the data model and the use of "simple and unsophisticated methods to find the desired probabilities or at least a good approximation to those probabilities. Thus approximate solutions were found to exact problems" (Conover [1]). These procedures became known collectively as *nonparametric* or *distribution-free* statistics (because few or no assumptions are made about the underlying distribution). They have several additional advantages:

1. They are easier and quicker to apply than many parametric methods.
2. The theory of many nonparametric methods is approachable with simple algebra and so is more readily understood by the scientist or technologist who uses them.
3. Because nonparametric methods are easily understood, the user is better able to judge the correctness of the use of a particular method.
4. The scientist is better able to develop his or her own nonparametric statistical methods, if the model is one not yet considered in the statistical literature.

Robust statistics provide an alternative approach to data analysis from that of nonparametric statistics by giving us a collection of theories dealing with approximate parametric models. It helps us to assess quantitatively the extent of deviations from the idealized assumptions of parametric statistics. Hampel et al. [2] list the main aims of robust statistics as follows.

DATA ANALYSIS

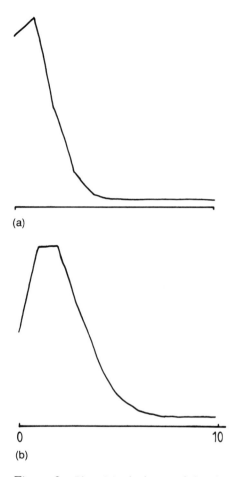

Figure 2 Plots (a) of a binomial distribution, (b) a Poisson distribution.

1. To describe the structure best fitting the bulk of the data
2. To identify deviating data points (outliers) or substructures for further treatment, if desired
3. To identify and give a warning about highly influential data points ("leverage points," see Sections 3.2 and 3.3)
4. To deal with unsuspected serial correlations or, more generally, with deviations from the assumed correlation structures.

The difference between nonparametric and robust statistical methods may be summarized as follows: Nonparametric statistics allows consideration of all possible models for the distribution of the experimental data, whereas robust statistics is closely related to parametric statistics and is concerned with describing the behavior of the majority of the data, identifying deviating points (outliers) or sets of such points and also highly influential points (points of high leverage that pull a regression in their direction). So, although some nonparametric methods are very robust, only some robust methods are nonparametric.

Mosteller and Tukey [3] give several examples of data analysis in which assumptions that one's data may fit a defined distribution may be invalid and analysis using standard parametric methods may lead to false confirmation or rejection of hypotheses. They make an effective case for exploring the data first, to try to understand their behavior, before homing in on a particular statistical test. Such exploratory methods have developed from the pioneering work of Tukey [4] in devising simple ways of examining data. Such methods may be *robust* in protecting the user from obtaining distorted conclusions from the data analysis. Unfortunately, such distortion will occur in parametric statistical analysis when even quite small deviations from the strict assumptions about the data distribution are found in experimental observations. They may also be *resistant* to distortion of conclusions about a set of observations resulting from the presence of unusual data values, whether these are genuine or arise from transcription errors. Exploratory methods make frequent and effective use of graphical, symbolic, or illustrative techniques for highlighting unusual data behavior or guiding one to the identification of general patterns in the data. They may also provide ways of diagnosing whether data need transformation before one attempts further statistical analysis [5].

Exploratory data analysis focuses one's attention onto four major areas [5]:

1. *Displays* reveal data behavior, both in their raw (unprocessed) form and as they progress through the various stages of statistical analysis.
2. Examination of *residuals* after particular analyses have been carried out often reveals useful features of the data, and the residuals may themselves be the subject of further detailed analyses.
3. Displays and the examination of residuals may reveal the need to *reexpress* the data, either to stabilize the variance or as a first step in nonlinear regression.
4. Exploratory methods have been designed to be *resistant* to undue influence of wild data values and to pay attention to the majority of the data.

Thus exploratory data analysis is a useful prelude to other forms of analysis, as a safeguard against faulty or inappropriate conclusions and as a guide and reviewing tool at each stage of data analysis. It should form an essential part of the toolkit for good statistical practice.

DATA ANALYSIS

Now let's have a look at various methods, their practical applications, and their limitations and advantages. No attempt will be made to be comprehensive because of the restrictions on the space available for this chapter, but the reader will be referred to key textbooks, monographs, papers, and software that explore these areas in much greater detail.

3.1 LIMITATIONS OF PARAMETRIC DATA ANALYSIS

When doing parametric statistical analysis of data, one assumes that the data are sampled from a distribution that has a well-defined form, such as the normal or Gaussian distribution (Figure 1) or the Poisson distribution (Figure 2). Following on from that assumption, one can specify the properties of the frequency distribution of the data by means of a small number of *parameters* such as the mean and standard deviation for the normal distribution. These parameters act as coefficients in the equations describing the distribution. If the data are reasonably described by such a well-defined distribution equation, it is usually a relatively straightforward matter to obtain estimates of these parameters from the data. Thus, the mean of a set of normally distributed data is the average (i.e., the arithmetic mean). The corresponding parameter for data following a lognormal distribution is the geometric mean.

A more general term for the parameter describing the middle value of a set of data is the *location* of the data distribution. The *spread* of the data is another useful general term. The spread of a normally distributed set of data is summarized by the standard deviation or variance.

Even simple statistical parameters such as the arithmetic mean and the standard deviation are potentially unreliable estimators of location and spread of the data, especially if the data are not truly Gaussian. Table 1 shows artificial data constructed from a Gaussian distribution and also the effect of distorting one of the numbers in the data on the mean and standard deviation. Table 2 illustrates this effect of outliers using real data [6] in estimating the precision or reproducibility of various sensors in replicate assays of potassium ion concentration, $[K^+]$, in whole blood.

Often, when measuring two or more variables, one is interested in analyzing the quantitative relationship between them. The process by which this relationship is investigated is known as *regression analysis*. In this process, we may try to fit a function $f(x)$ of one variable, x, to another variable, y, such that, for example,

$$y = f(x)$$

Other purposes of regression include (1) the setting aside of variables that are measured but do not contribute to the variation in y, (2) analysis of the causes

Table 1A Effect of Presence of an Outlier on Mean and Standard Deviation

Sample of 10 from a Gaussian distribution of mean = 10 and SD = 1	Same sample contaminated with one observation from Gaussian with mean = 10 and SD = 3
9.6615	9.6615
10.0543	3.6269
9.0088	9.0088
10.1617	10.1617
10.5272	10.5272
10.7719	10.7719
8.1165	8.1165
8.7944	8.7974
8.2926	8.2926
11.0888	11.0888
Mean of sample = 9.648	Mean of sample = 9.005
SD of sample = 1.047	SD of sample = 2.155

Table 1B Data of Table 1A Ranked in Increasing Order

Rank	Gaussian sample	Contaminated Gaussian sample
1	8.1165	3.6269
2	8.2926	8.1165
3	8.7974	8.2926
4	9.0088	8.7974
5	9.6615	9.0088
6	10.0543	9.6615
7	10.1617	10.1617
8	10.5272	10.5272
9	10.7719	10.7719
10	11.0888	11.0888
Median	9.8579	9.3351

of natural or experimental phenomena, (3) prediction of the effects of changes in x on the magnitude of y, (4) calibration of an assay or measurement system, and (5) performance evaluation of a measurement system and/or its comparison with established or standard systems.

If one assumes that the y variable follows a normal distribution, and y is a linear function of x, then the method of ordinary least squares developed by Gauss in the early part of the nineteenth century (e.g., see Deming [7]) is the

DATA ANALYSIS

Table 2 The Effect of Outliers of Replicate Range (mM) on Estimates of Imprecision of Assays of Whole Blood for [K^+] Using ISFETs and ISEs

Sensor	CV with outliers, %	CV without outliers, %
Thorn EMI ISFET	4.00	2.88
Corning 902	3.79	2.53
Chempro 500	10.68	8.07

favored approach to regression analysis. Ordinary least squares (OLS) regression is achieved by trying to fit a relationship

$$Y_i = \beta_0 + \beta_1 X_i + e_i$$

by minimizing the sum of the squares of the deviations from the line

$$Y_i = \beta_0 + \beta_1 X_i$$

The OLS solution to this minimization is given by the following equation for the slope of the fitted straight line:

$$b_1 = \frac{\sum_i^n X_i Y_i - [(\sum_i^n X_i)(\sum_i^n Y_i)]/n}{\sum_i^n X_i^2 - (\sum_i^n X_i)^2/n} = \frac{\sum_i^n (X_i - \overline{X})(Y_i - \overline{Y})}{\sum_i^n (X_i - \overline{X})^2}$$

and the intercept is given by

$$b_0 = \overline{Y} - b_1 \overline{X}$$

Because the equation for the slope involves squared terms, it tends to exaggerate the influence of points that deviate strongly from the bulk of the data. As a result, it is possible for the OLS fitting process to produce a fitted line whose position is determined just as much by this deviant point as by the bulk of the data, as may be seen in Figure 3.

The tolerance of a regression method to the presence of such deviant points can be described by means of the breakdown bound proposed in 1971 by Hampel [8], which is defined as follows: The *breakdown bound* for the fitting of n pairs of y versus x data is k/n, k being the largest number of data points replaceable by arbitrary values yet not greatly changing the fitted line. The maximum value of the breakdown bound is 1/2, because a fitting procedure must at least represent the majority of the data and so one cannot tolerate more than 50% of the data going arbitrarily wild. The least squares regression method of fitting a line to data has a breakdown bound of zero; that is, it is intolerant of all outliers for the reasons already discussed.

Figure 4 illustrates some of the problems of parametric testing, as found in the use of regression for the comparison of two assay methods or sensor systems.

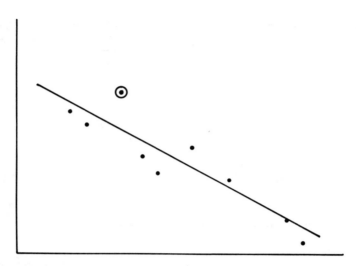

Figure 3 Bivariate scatterplot with the ordinary least squares regression line showing an outlier point (circled) and its pulling influence on the position of the regression line.

Paired measurements of sodium ion concentration, [Na$^+$], in whole blood by an ISE-based instrument and by an ISFET [9] are plotted as a two-dimensional scatterplot, and various regression lines are drawn on the graph. The ordinary least squares (OLS) line is calculated using a standard procedure available in many statistical software packages suitable for use on a PC. In such an application, a major flaw of OLS regression is the assumption that there is no error in the x variable (the reference assay method or instrument, in this case). Clearly, this assumption is not valid in this situation in which method comparisons are of interest, and recommendations have been made in the clinical chemistry literature [10,11] about the use of a modified OLS method introduced by Deming [7] (see also Mandel [12] for a description of useful algorithms for implementation of Deming's method). In this approach, the equation for the regression coefficient (slope) with a correction for the errors in the x variable is

$$\beta' = \{\phi w - u + [(u-\phi w)^2 + 4\phi p^2]^{1/2}\}/2\phi p$$

where

$u = N \Sigma (x_i - x)^2$
$w = N \Sigma_i (y_i - y)^2$
$p = N \Sigma_i (x_i - x)(y_i - y)$
$\phi = $ (variance of x)/(variance of y)

DATA ANALYSIS

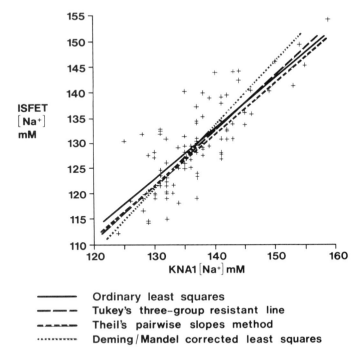

Figure 4 Various regression plots of measurements of [Na$^+$] in whole blood using a Thorn EMI sodium ion selective field effect transistor (ISFET) compared with measurements on the same specimens made using a standard ion-selective clinical stat analyzer, the Radiometer KNA1.

The variances may be determined experimentally—for example, by replication of measurements.

The Deming–Mandel regression line for the ISFET/ISE comparison is also shown in Figure 4. Unfortunately, this method suffers from the same major limitation as the OLS method already mentioned above, namely, that various types of outlier points positioned away from the majority may have a very strong influence on the position of the calculated regression line. This is because OLS regression and its variants all use the squares of the deviations of each x and y value from their respective means in calculating the regression parameters. Thus, any large deviation exerts an influence proportional to the square of that deviation, which exaggerates its influence on the regression. As with OLS regression itself, Deming–Mandel regression has a breakdown bound of zero.

The results of using two alternative approaches to calculating the regression parameters, which are nonparametric and robust, are also shown in Figure 4. (These will be discussed in detail in Section 3.2 under the heading Nonparametric Approaches to Regression and in Section 3.3 under the heading Resistant Regression.) The actual regression parameters (slope and intercept) are listed in Table 3A. It would appear from examining Tables 3A and 3B and Figure 4 that the Deming–Mandel method overcorrects the OLS result for this particular set of data. The OLS results in this case compare favorably with those from the nonparametric methods. However, if one's aim were to estimate bias between the methods, the nonparametric regression estimates would be likely to be more reliable, as they account directly for errors in both variables and do not excessively weight influential or outlier points.

Influential or outlier points, identified by appropriate methods discussed in various sections of this chapter, should not merely be considered as candidates for rejection. That is not really the purpose of the identification processes. Apart from the important process of screening out data arising from errors of transcription or transmission or genuine experimental mistakes or faulty equipment, the

Table 3A Comparison of Regression Parameters Obtained Using Various Techniques of Regression Analysis for the Assay of Whole Blood [Na^+] Using Thorn EMI Microsensors ISFET and Radiometer KNA1

Regression method	Intercept	Slope
Ordinary least squares	−8.2	1.007
Deming–Mandel Least Squares	−43.0	1.261
Tukey's three-group resistant line	−19.5	1.083
Theil's pairwise slopes	−12.6	1.031

Table 3B Regression Fitted Values Calculated from Regression Parameters of Table 3A

Regression technique	KNA1 [Na^+] value	Predicted (fitted) ISFET [Na^+] value
Ordinary least squares	125	117.7
	150	142.9
Deming–Mandel least squares	125	114.7
	150	146.2
Tukey's three-group resistant line	125	115.3
	150	143.0
Theil's pairwise slopes	125	116.3
	150	142.1

DATA ANALYSIS

purpose is to use the highlighted data as a starting point for reexamining hypotheses about the systems and/or samples being measured and as a stimulus to the process of formulating new hypotheses. This is both a major philosophical difference and a practical difference from the approach adopted when parametric methods were the dominant tools and outliers were regarded dismissively as mere nuisance data. This very practical approach to outliers is at the very heart of late twentieth century statistics and chemometrics.

The rest of this chapter focuses on simple approaches to various needs in data analysis using the nonparametric, exploratory, and robust methods and a variety of techniques, some of which can be easily applied with pencil and paper alone or just with a simple electronic calculator, while others can be implemented with spreadsheets or statistical software packages or via readily available program listings in the literature. Most methods will be illustrated with a worked example as well as with problems that the reader may try.

3.2 NONPARAMETRIC METHODS

When a series of measurements are made on a single specimen or a single variable (e.g., $[Na^+]$) on a series of blood samples, it is useful to be able to estimate an average value (location) of the variable measured and how the measurements of the variable are spread. The median is a better estimate of the location than the arithmetic mean and has several desirable properties. It is resistant to distortion by the presence of extreme values in the data, and it is a more representative measure of location when data are sampled from unsymmetrical distributions. The median is the middle ranked value of a set of observations (when there are an odd number of them) or the mean of the middle two ranked observations (when there are an even number of them). The ranking of the data may be in either ascending or descending order. For example, the set of numbers

1.3, 1.5, 0.8, 0.6, 3.2, 5.1, 2.4, 9.0

can be ordered as follows:

Rank	Ascending order	Descending order
1	0.6	9.0
2	0.8	5.1
3	1.3	3.2
4	1.5	2.4
5	2.4	1.5
6	3.2	1.3
7	5.1	0.8
8	9.0	0.6

Testing Hypotheses About the Median

One may wish to test whether a particular set of soil specimens has an average concentration of a plant nutrient (e.g., phosphate) in the desired range for adequate growth of a crop. The parametric approach to this problem would be to apply the Student t-test in the form

$$t = \frac{\bar{x} - \mu}{(s/n)^{-1/2}}$$

where μ is the desired mean soil phosphate concentration, $\bar{x}$ is the arithmetic mean of the specimens, s is their standard deviation, and n is the number of specimens. If one were confident that the distribution of phosphate was Gaussian and that wild values were unlikely, then this test would be appropriate. But in practice one may not be confident, perhaps because it would cost too much to analyze a sufficiently large set of specimens to establish such confidence. Then either of the simple nonparametric methods known as the *sign test* and the *Wilcoxon signed rank test* is a useful alternative to the Student t-test. These tests differ in their applicability and also in their efficiency (as discussed below).

For Wilcoxon's test, the sample of data must be drawn from a symmetrical distribution (this is a much more relaxed requirement than for Student's t-test, for which the sample must be from a Gaussian distribution). The sign test has no restriction on the form of the distribution from which the sample is drawn and is thus more genuinely distribution-free. The efficiency of a test relates to the size of sample needed to reach a conclusion with a particular level of confidence. A lower efficiency for a test implies that it requires a larger number of observations in the data set to achieve the same level of confidence as that achieved by a higher efficiency test. In this respect, the sign test is less efficient than the Wilcoxon signed ranks test with data from symmetrical distributions. Nonetheless, if we are concerned that the sample of measurements is not from a symmetrical distribution, or if it contains outliers, then it may be the more appropriate test to use.

The differences between these two tests and the Student t-test are illustrated below using three sets of data from a Gaussian, a symmetrical (uniform), and an unsymmetrical (lognormal) distribution, respectively (see Table 4A).

For single samples of data such as those in each of the columns in, Table 4A, we can set up a hypothesis that the sample median or mean is a particular value. For example, for columns 1 and 2 the hypothesized median could be set at 10, and for column 3 it could be set at 3.

With the sign test, we proceed by taking the hypothesized median value from each of the values in the column and then counting the number of positive or negative differences (see Table 4B). For the data in column 1, there are nine positive and 11 negative signs. The test is based on the binomial distribution

DATA ANALYSIS

Table 4A Three Samples of Data

	Distribution Type		
Row	Sample 1 Gaussian	Sample 2 Uniform	Sample 3 Lognormal
1	11.0417	9.9461	3.07649
2	10.7695	13.2661	2.64697
3	11.3381	8.2614	3.00518
4	10.2083	12.6805	3.01085
5	9.0484	5.0653	2.54949
6	9.9653	13.1637	2.89295
7	9.4190	5.4686	2.48317
8	8.5130	13.5812	2.85827
9	11.8342	7.6533	2.86570
10	9.2183	6.6569	2.95362
11	8.6944	12.1117	3.06081
12	7.1678	13.9577	3.22360
13	11.6098	14.7153	2.91975
14	6.7009	9.4111	2.71892
15	12.2758	6.3892	2.73682
16	9.2980	8.6551	2.42275
17	10.3197	11.8915	2.99465
18	8.3014	6.4316	2.95739
19	12.8891	13.9444	2.77740
20	5.3143	13.0559	2.58356

with an equal chance of plus or minus signs; that is, for any given difference there should be an equal chance (probability) of that difference being positive or negative. This is termed a binomial distribution with $p = 1/2$. We now need to know, for any given set of such differences, what the probability is of obtaining the observed set of differences. This can be calculated for any given sample size, and tables of such probabilities are to be found in many statistics textbooks. Checking in a table of binomial probabilities for a sample of 20 with $p = 1/2$ (see Table 5), we find that for nine plus signs, the probability value is 0.4119. Now we are testing whether the median is either above or below 10, so we have a two-sided test (testing both alternatives to the null hypothesis), and our p-value must be doubled to 0.8238. This value is large, so we may conclude that the median that we guessed in our original hypothesis was a very reasonable estimate. If we had had a probability value of 0.05 or less, we could have concluded that the set of differences was unlikely to provide support for the hypothesis that the two tests were comparable.

Table 4B Calculations for Sign, Wilcoxon Signed Rank, and Student's t Tests Illustrated for Sample 1

Row	Value x_1	Difference from actual mean x_m, $x_1 - x_m$	Difference from hypothesized median x_M, $x_1 - x_M$	Ranks of differences from median	Signed ranks of differences from median
1	11.0417	+1.3457	+1.0417	9	+9
2	10.7695	+1.0735	+0.7695	6	+6
3	11.3381	+1.6421	+1.3381	11	+11
4	10.2083	+0.5123	+0.2083	2	+2
5	9.0484	−0.6476	−0.9516	8	−8
6	9.9653	+0.2693	−0.0347	1	−1
7	9.4190	−0.2770	−0.5810	4	−4
8	8.5130	−1.1830	−1.4870	12	−12
9	11.8342	+2.1382	+1.8342	15	+15
10	9.2183	−0.4777	−0.7817	7	−7
11	8.6944	−1.0016	−1.3056	10	−10
12	7.1678	−2.5282	−2.8322	17	−17
13	11.6098	+1.9138	+1.6098	13	+13
14	6.7009	−2.9951	−3.2991	19	−19
15	12.2758	+2.5798	+2.2758	16	+16
16	9.2980	−0.3980	−0.7020	5	−5
17	10.3197	+0.6237	+0.3197	3	+3
18	8.3014	−1.3946	−1.6986	14	−14
19	12.8891	+3.1931	+2.8891	18	+18
20	5.3143	−4.3817	−4.6857	20	−20

We could set a one-sided hypothesis for the data from column 3—for example, that the median was greater than 3. The reader is invited to try the sign test on these data. The p value you should get is 0.0207 from the table of binomial probabilities (Table 5). This is well below 0.05, and the conclusion reached is that this is a poor estimate. For columns 1 and 2, the samples are drawn from Gaussian and uniform distributions, respectively, so the mean and median coincide, and a test for the median is also a test for the mean. The data in column 3 are drawn from a lognormal distribution, in which the mean does not coincide with the median, so then the test is only a median test.

The test can be modified [13] to test any quantile or percentile, not just the median (50th percentile). Because this test is so simple to do, it does not need a computer and complex software for small sets of observations.

Wilcoxon's signed rank rest [14] can be applied to the same data by ordering the data in the sample and assigning a rank to each observation. To

DATA ANALYSIS

Table 5 Cumulative Binomial Probabilities for $n = 20$ and $p = 1/2$

Left S	P	Right S
0	0.0000	20
1	0.0000	19
2	0.0002	18
3	0.0013	17
4	0.0059	16
5	0.0207	15
6	0.0577	14
7	0.1316	13
8	0.2517	12
9	0.4119	11
10	0.5881	10

each such rank is then attached the plus or minus sign denoting whether that particular observation is above or below the hypothetical median being tested, as illustrated in Table 4B for the data in Table 4A, column 1. In Table 4B, column 1 corresponds to column 1 of Table 4A; column 3 shows the differences from the hypothesized median, and these are the differences we would use in testing whether the proposed median is reasonably close to the actual median. Column 4 shows the ranks of the differences, with the ranking done *ignoring the sign of the difference*. The next step is shown in column 5, in which we attach the sign of the difference to the corresponding rank. Then either the total of the positive (T_+) or of the negative (T_-) signed ranks is calculated. The equation relating the two totals is

$$T_+ + T_- = n(n+1)/2$$

where n is the number of observations. The larger total is generally used as the test statistic [14].

Minitab uses the Hodges—Lehmann estimator [15] to perform the Wilcoxon signed ranks test, and according to the Minitab reference manual [16] this is an algebraically equivalent form of the test. Conover [17] gives an alternative test statistic for the signed ranks test, which is especially useful when several observations have the same rank (tied ranks).

Calculating Confidence Limits on the Median

The sign test can also be used to calculate confidence limits for the actual median of the sample. The observations are ordered, and our task is to find the positions in the ordered list that correspond to the desired confidence limits. If we look at Table 5 we can see that the p value is 0.0207 at ranks 5 and 15 and

0.0577 at ranks 6 and 14. These give two-sided p values of 0.0414 and 0.1154, so unless we have a means of interpolating to any arbitrary value in between these exact confidence limits, we are restricted to the exact limits. Thus, for example, for $p=0.0414$, the ranks used to estimate the limits are one rank in from the estimated ranks 6 and 14 mentioned above. The observations corresponding to these in the three columns in Table 4A are:

    Column 1:   8.694 and 11.042,   median = 9.69
    Column 2:   7.653 and 13.164,   median = 10.92
    Column 3:   2.719 and 2.995,   median = 2.879

It can be seen that the sample from the uniform distribution (column 2), although having the same mean and standard deviation as the Gaussian sample (column 1) has a much wider confidence interval.

Some statistical software packages, such as Minitab, will perform the confidence limit calculations to give the 95% limits by a nonlinear interpolation as well as exact limits on either side of the 95% limit. The interpolated (95%) confidence limits are

    Column 1:   8.78,   10.98
    Column 2:   7.80,   13.14
    Column 3:   2.723,   2.986

With Minitab, the Hodges–Lehmann estimator is used to calculate the confidence interval based on the signed ranks test [16]. The means of every pair of the n observations in the sample, including the pairs of observations with themselves, are calculated [a total of $n(n+1)/2$ pairs]. Using a table of critical values for the Wilcoxon signed rank test [18] for the samples of 20 numbers in Table 4A, we find that for the two-sided test the 52nd highest and lowest ranked of these means are the 5% confidence boundaries of these samples. Minitab computes these with the command WINTERVAL confidence = k for c,..,c. For the three sets of data in Table 4A, the 95% confidence intervals are as follows.

	Estimated median	95% Confidence interval
Col. 1	9.773	8.800, 10.662
Col. 2	10.23	8.836, 12.286
Col. 3	2.848	2.735, 2.956

These intervals are narrower than those calculated from the sign test because more information from the sample is used in the estimation. This computation does, however, assume that the sample distribution is symmetrical and does tend to get distorted by the presence of extreme outliers. In that sense, the interval

DATA ANALYSIS

found from the sign test is more conservative and probably more resistant. If little is known about the distribution from which the sample of observations is taken, or if it is suspected to be non-Gaussian, then these tests are likely to be much safer than Student's t-test. For comparison, the confidence intervals obtained using t are as follows.

	Mean	SD	95% Confidence interval
Col. 1	9.696	1.937	8.790, 10.603
Col. 2	10.315	3.261	8.789, 11.842
Col. 3	2.8369	0.2161	2.7358, 2.9381

The same kinds of calculations can be performed using paired observations, as nonparametric alternatives to the paired t-test. This can be very useful in method comparisons. Differences between the paired observations are used in exactly the same way as for the single-sample cases described above, with the rider that zero differences are not used in either the sign or Wilcoxon signed rank tests.

Tests on Paired Observations

Paired data can be tested in the same way as data from single samples described in the previous two sections, because the comparisons between the two sets of measurements are made using the differences between pairs. Table 6 shows a set of paired data comparing plasma chloride assays performed by coulometric and potentiometric methods. The reader is invited to perform the sign, Wilcoxon, and Student's t tests on these data to test whether there is any significant difference between the methods and then, if appropriate, to find the confidence interval on any mean or median difference obtained. Some tips are needed for the Wilcoxon test:

1. The differences are ranked according to their absolute value (ignoring the sign).
2. The ranks are given the same sign as the corresponding difference.
3. Zero differences are ignored; ranking is done only with nonzero differences.
4. Differences with the same absolute value ("tied" differences) are given the same average rank.

Comparing Sets of Unpaired Observations

When comparing two sets of unpaired observations, the nonparametric analog of the unpaired t-test is the Mann–Whitney–Wilcoxon test. Table 7A lists two

Table 6 Human Plasma [Cl⁻] Assays by Corning 925 Chloride Analyzer and Ektachem DT60/DTE Chloride Slides

Plasma spec. number	[Cl$^-$], mM	
	Corning	Ektachem
1	99	98
2	99	101
3	104	105
4	98	94
5	98	93
6	103	99
7	108	101
8	122	120
9	122	119
10	95	91
11	98	92
12	103	99
13	93	89
14	107	100
15	84	84
16	95	102
17	89	92
18	107	107
19	113	113
20	109	111
21	108	108
22	98	97
23	114	113
24	115	121
25	110	113
26	104	106
27	111	115
28	103	98
29	115	117
30	93	87

unpaired samples of data, and Table 7B illustrates the method of processing the data in this test. First, the data from each sample are combined and placed in order in a single list, the sample from which each observation originated is indicated by the numbers being italic or nonitalic, and the rank for each is listed and similarly identified. The objective of the test is to determine whether the medians of the two samples are the same. It does not matter whether samples

DATA ANALYSIS

Table 7A Unpaired Data for Analysis by Mann–Whitney–Wilcoxon Test

Sample of 10 from a Gaussian distribution with mean = 10 and SD = 1:
9.3578	9.7058	11.1556	10.2497	9.9609	10.3149	10.1752
10.0441	10.1063	10.4246				

Sample of 13 from a Gaussian distribution with mean = 11 and SD = 1:
10.4915	11.8330	11.2560	12.5812	9.6577	12.0960	10.0453
11.4482	12.5977	12.4735	11.2662	8.2696	10.5555	

Table 7B Ordering of Data for Mann–Whitney–Wilcoxon Test

Rank	1	2	3	4	5	6	7
Data	8.2696	9.3758	9.6577	9.7058	9.9609	10.0441	10.0453
Rank	8	9	10	11	12	13	14
Data	10.1063	10.1752	10.2497	10.3149	10.4246	10.4915	10.5555
Rank	15	16	17	18	19	20	21
Data	11.1556	11.2560	11.2662	11.4482	11.8330	12.0960	12.4753
Rank	22	23					
Data	12.5812	12.5977					

have equal numbers of observations or not. The sum of the ranks of the sample with the smaller number of observations is used as the test statistic.

Sprent [19], Conover [20], and Gibbons [21] give tables enabling one to perform this test using the rank sum statistic. An alternative form of the test uses a statistic, U, defined as [22]

$$U_m = S_m - \frac{m(m+1)}{2}$$

where S_m is the sum of the ranks of the sample with the smaller number of observations, m.

Comparing Several Sets of Observations, One-Way Analysis of Variance

The Mann–Whitney–Wilcoxon test for two independent samples can be naturally extended to k such samples, enabling one to perform a one-way analysis of variance (ANOVA) by ranks. This extension was derived by Kruskal and Wallis and, according to Conover [23], is more reliable than standard "Gaussian theory" analysis of variance in the presence of outliers. The ranks of all the data considered collectively are obtained and allocated back to the original columns, and sums of the ranks in each column are used in calculating the test statistic (Gibbons [24]):

$$H = \frac{12}{N(N+1)} \sum_{j=1}^{j=k} \frac{[R_j - n_j(N+1)/2]^2}{n_j}$$

where N is the total number of observations, n_j is the number of observations in the jth column, and $j = 1, \ldots, k$, and R_j is the sum of the ranks in that jth column. If data are tied (equal observations occurring in one or more columns), the middle rank of the set of equal observations is given to each one. For example, if four observations are the same starting at rank 5, then the rank positions occupied by these observations are 5, 6, 7, and 8, but we do not know how to allocate those ranks to the observations. The solution is to take the average of the ranks, in this case 6.5, and allocate that rank to each of the tied (equal) observations. If there are extensive ties in the set of data, a correction to the test statistic H is obtained by dividing H by

$$1 - \frac{\Sigma(u^3 - u)}{N(N^2 - 1)}$$

where u is the number of all observations, in all columns, that are tied for any given rank and the sum is over all sets of tied ranks. The test statistic H may be compared with the χ^2 values for k-1 degrees of freedom. If H exceeds the χ^2 value, this shows that there is a significant difference between columns; then we need a multiple-comparisons procedure to identify the extent of the column differences. Conover [23], Gibbons [24], and Sprent [25] all give descriptions of such procedures, which enable one also to assign a significance to any detected difference. The Kruskal–Wallis test procedure is illustrated in Table 8.

One may not necessarily be interested only in differences in the average value (location) of sets of observations. Differences in the spread of each set of data or in the shapes of the sampled data distributions can be detected by various nonparametric procedures. The squared-ranks test can be used to check whether or not two or more sets of observations have equal spreads. The interested reader is referred to Conover [26] or Sprent [27] for detailed descriptions of the test.

Tests Using Cumulative Frequency Distributions

The Kolmogorov–Smirnov one-sample test can be used to compare a sample of data with any given data distribution, for example, normal or lognormal. The

DATA ANALYSIS

Table 8 Kruskal–Wallis One-Way Analysis of Variance by Ranks of Data from Table 4B

(A) Ranks of data in Table 4B

	Column 1	Column 2	Column 3
	45	40	19
	44	56	5
	46	30	16
	42	52	17
	35	21	3
	41	55	11
	39	23	2
	32	57	9
	48	29	10
	36	26	13
	34	50	18
	28	59	20
	47	60	12
	27	38	6
	51	24	7
	37	33	1
	43	49	15
	31	25	14
	53	58	8
	22	54	4
Totals	781	839	210

(B) Test statistic calculation

$$H = \frac{12}{60(60+1)} \left(\frac{781^2 + 839^2 + 210^2}{20} \right) - 3(60+1)$$

Compare with χ^2 for $k-1=2$ degrees of freedom.
χ^2 for $p=0.001$ is $13.82 \ll H$, so there is a significant difference between columns

two-sample version of this test can be used to compare the distributions of two sets of data. In either case, one uses the cumulative frequency distributions, and the test statistic is the largest difference between the two distributions being compared. Table 9 and Figure 5 illustrate the procedure using data from a comparison of the two [Cl$^-$] assay methods from Table 6. This type of test is a general test for any major difference between samples, whether it is location, spread, or distribution shape. For further reading on this useful test, the reader is referred to Conover [28], Gibbons [29], or Sprent [30].

Table 9 Kolmogorov–Smirnov Two-Sided Two-Sample Test of [Cl$^-$] Assay Data from Table 6

Corning [Cl$^-$], mM	Freq.	Cum. freq.	Ektachem [Cl$^-$], mM	Freq.	Cum. freq.
84	0.03333	0.03333	84	0.03333	0.03333
89	0.03333	0.06666	87	0.03333	0.06666
93	0.06666	0.13333	89	0.03333	0.1
95	0.06666	0.2	91	0.03333	0.13333
98	0.13333	0.33333	92	0.06666	0.2
99	0.06666	0.4	93	0.03333	0.23333
103	0.1	0.5	94	0.03333	0.26666
104	0.06666	0.56666	97	0.03333	0.3
107	0.06666	0.63333	98	0.06666	0.36666
108	0.06666	0.7	99	0.06666	0.43333
109	0.03333	0.73333	100	0.03333	0.46666
110	0.06666	0.8	101	0.06666	0.53333
113	0.03333	0.83333	102	0.03333	0.56666
114	0.03333	0.86666	105	0.03333	0.6
115	0.06666	0.93333	106	0.03333	0.63333
122	0.06666	1.0	107	0.03333	0.66666
			108	0.03333	0.7
			111	0.03333	0.73333
			113	0.1	0.83333
			115	0.03333	0.86666
			117	0.03333	0.9
			119	0.03333	0.93333
			120	0.03333	0.96666
			121	0.03333	1.0

Nonparametric Approaches to Regression

Ordinary least squares (OLS) regression assumes that the data being evaluated and belonging to the y variable are normally distributed and those belonging to the x variable have little or no error. In reality with chemical/physical measurements, the two data sets being compared by regression both have errors, and Deming [7] and Mandel [12] devised procedures to allow for this in OLS regression. As pointed out earlier in the chapter, however, the Deming–Mandel procedure and the OLS method have a breakdown bound of zero when they encounter outlying data, and both are inappropriate. So we need to look for a method with a reasonable breakdown bound to give us some protection.

A useful nonparametric regression method with a breakdown bound of 0.29 and hence a considerable degree of robustness against wild data was devised

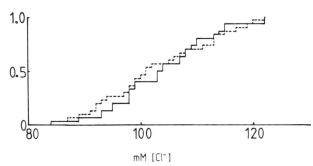

Figure 5 Cumulative frequency distributions for [Cl⁻] assay in human plasma as measured coulometrically by the Corning 925 Chloride Analyser (full line) and potentiometrically by the Kodak Ektachem DT60/DTE Analyser (broken line) as used for the Kolmogorov–Smirnov two-sided two-sample test.

by Theil and is described in detail by Sprent [31]. Theil's so-called complete method involves calculating all possible slopes between pairs of (x,y) values. In Minitab this is achieved using the WSLOPES command. The median of all these slopes is the regression coefficient. For each (x,y) pair, we can use this estimated regression coefficient to estimate a corresponding value for the intercept from the equation

$$a_n = y_n - b_{med}x_n$$

The median of these intercepts is used as the constant term in the regression equation.

Various authors have described ways of calculating a confidence interval around the regression coefficient (see Sprent [31]), but in chemical applications it would be more useful to have a confidence envelope around the regression line. Suitable methods are discussed in detail by Maritz [32]. Examples of other methods are to be found in Thompson [33] and Lancaster and Quade [34]. None of these methods is available in any commercial software. Unfortunately, this problem of confidence envelopes seems to be regarded by many statisticians as relatively unimportant despite its obvious practical value in calibration and prediction. One of the major disincentives may have been the relatively large computing power needed, although with the newer generations of personal computers using 80386 or 80486 chips with the corresponding math coprocessors, computing power should no longer be a major problem.

Nonparametric Two-Way Analysis of Variance

The examination of two-way randomized block designs can be carried out using a ranks method introduced by Friedman. Several versions of the test statistic have been devised, and these are critically assessed by Conover [35] and Sprent [36], who also discuss multiple comparisons within this type of two-way ANOVA. The major advantage of this procedure is its resistance to violations of the normality assumptions; its principal disadvantage is the inability to cope with empty cells in the block design. Alternative approaches are dealt with briefly in Section 3.3 under the heading Exploratory Methods for Two-way Analysis of Variance. These methods can cope easily with missing values and exhibit useful resistance to wild data.

3.3 EXPLORATORY DATA ANALYSIS

The Use of Scatterplots

Among the simplest exploratory techniques for univariate or bivariate data are various forms of scatterplots. These enable us to reveal patterns or explore differences, for example. Figure 6 shows an example of a vertical scatterplot of the ranges of duplicate measurements of $[K^+]$ in whole blood as measured by a variety of ion-selective sensors. One of the sensors tested is seen to be very much less reproducible than the others, having much more scatter of range values. This kind of plot is especially helpful in examining the contribution of discrepant range values in assessing precision. For example, outliers may distort the combined variance calculated from such a set of ranges as in the cases of the Chempro 500, the ISFET, and the Corning 902 in Figure 6. Removal of one or two outlying values from the calculation can have quite a marked effect as shown in Table 3. Figure 7 shows the scatterplots of range from several series of duplicate assays of whole-blood $[Na^+]$ by ISFETs, which serves to illustrate the often ignored variation from one such series to another and that outliers do not always occur [37].

The scatterplot plays a major role in preventing the facile misinterpretation of bald number summaries, such as correlation coefficients, as is dramatically illustrated in Figures 8a and 8b, which show eight plots of sets of data, all with correlation coefficients of 0.7. These demonstrate clearly the ease with which one can be misled. One is prompted to recall the abuses and misuses of statistics that have in the past led to such cynical remarks as that of Mark Twain [38]: "There are three kinds of lies: lies, damned lies and statistics."

DATA ANALYSIS

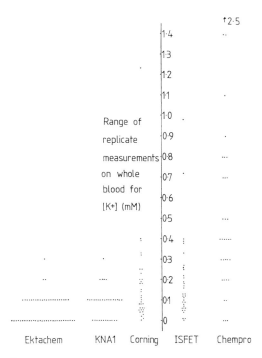

Figure 6 Vertical scatterplots of the ranges between duplicate assays for whole blood [K$^+$] measured using a Thorn EMI potassium ISFET, the Corning 902 and Radiometer KNA1 sodium/potassium analyzer, Kodak the Ektachem DT60/DTE analyzer and Chempro Ion Profile sensor cards. Each symbol represents the range for a single blood specimen.

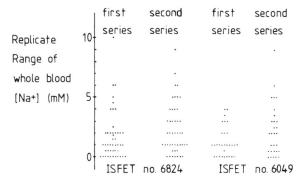

Figure 7 Vertical scatterplots of the ranges of duplicate assays for whole blood [Na$^+$] measured using various Thorn EMI sodium ISFETs.

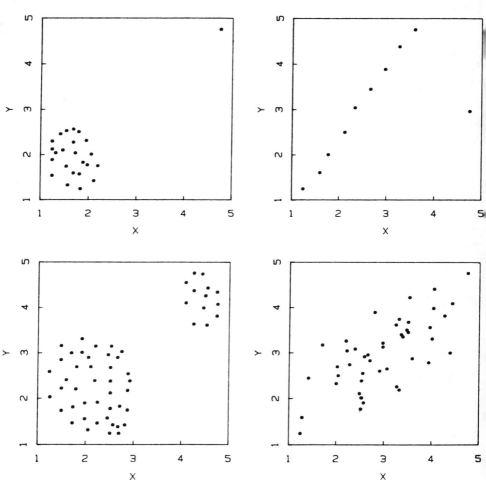

Figure 8 Various bivariate scatterplots, each of data having a correlation of 0.7. (Reproduced by permission of Wadsworth and Brooks/Cole Advanced Books and Software, Pacific Grove, CA 93950.)

Stem-and-Leaf Displays

The stem-and-leaf display [39] is closely related to the histogram but provides the data analyst with a more detailed picture. It can serve also as a useful aid to ranking by hand for simple calculations on data structure (to be discussed in the next subsection). Figure 9 illustrates the construction of such a display. The

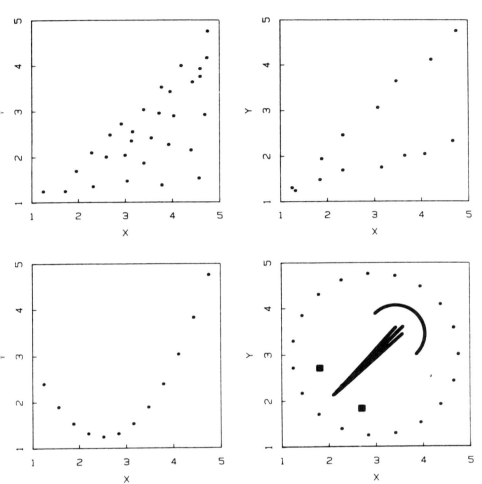

Figure 8b

stem is made from the first one or two digits of each of the observations, and sometimes, as in this display, the extent of numbers starting with these first digits is sufficiently large to warrant a split of the group. The leaves are the next digit of each observation, so that, for example, with the data displayed here in Figure 9, at the first stem position down marked 13 there are five zero values; each represents an observation value 130. Then there is a single 1, representing 131; five 2s for five observations of 132; and so on.

```
N=98         LEAF UNIT=1.0mM


    2    12   24
   12    12   6677888999
   36    13   000001222223333344444444
  (18)   13   555556667788888999
   44    14   00000011111111222222223333444
   16    14   556667889
    7    15   11224
    2    15   88
```

Figure 9 Stem-and-leaf display.

Next to the display is a cumulative count of the number of observations in the display counting inwards from each end of the display. For example, by the end of the lower row marked 12, there are 12 observations accumulated, and by the end of the next row down, 36 have been counted. The row in which the median is present is marked with a number in parentheses; this is just the count of digits in that row. Figure 10 illustrates the idea of using stem-and-leaf displays in a "back-to-back" mode in order to compare distributions of data. It shows the flame photometer assays of [Na] in blood plasma (from Figure 9) compared to whole blood [Na^+] assays by an ISFET on the same specimens [37]. From this, one can see that the distributions are offset from one another but are not markedly different in spread or shape.

Letter Values and Letter Value Displays

Simple resistant summaries of data, which are very useful in exploring data, are the "letter values," so called because Mosteller and Tukey [40] tagged each value with the first letter of its name. The most familiar of these is the median, which was tagged M and is the middle ranked value in an odd-numbered set of observations or the mean of the middle two ranked values in an even-numbered set. This is much less affected by the presence of a single wild data value than the mean.

We can use letter values to describe the profile of a data distribution spreading progressively outwards in both directions from the median. As we move outwards from the median in a ranked list of data, the positions of the

```
N=98        LEAF UNIT=1.0

    2                              42 11
   14                   999999998865 11
   29                   443333211110000 12 24                              2
  (24)  999999988888877776666655 12 6677888999                            12
   45   4444333332222211111110000 13 000001222223333344444444             36
   21                      9998765 13 555556667788888999                 (18)
   14                      444210000 14 000000111111122222223333444       44
    5                           9665 14 556667889                         16
    1                              4 15 11224                              7
                                     15 88                                 2

                NaISFET                    FLAME PHOTOMETRY
```

Figure 10 Back-to-back stem-and-leaf display.

letter values can be found in relation to the positions of the previous values from the term [41]

$$\frac{(\text{Previous depth}) + 1}{2}$$

The *depth* is a measure of the position in from either end of the ranked list and is found as follows. First, we sort the data into ascending order of size. Now it is possible to rank any number in this ordered set by counting either from the smallest upwards or from the largest downwards, as already discussed at the beginning of Section 3.2. The first kind of rank is termed the upward rank, and the second is the downward rank. For any data value we have

Upward rank + downward rank = $n + 1$

where n is the number of observations. We can define the *depth* of a data value in an ordered set as the smaller of its upward and downward ranks.

After the median, the next useful resistant letter value is the *fourth* (which is tagged with the letter F) [41], alternatively called the *hinge* (tagged H) [40], a form of quartile defined by

$$\text{Depth of fourth} = \frac{(\text{depth of median}) + 1}{2}$$

The rule to be used when applying this equation is to drop any fraction from the depth of the median before using it in the equation; then the depth of the fourth can either be an integer or an integer plus 1/2 [41].

We can make a five-number summary [41] of the data from the median, the fourth, and the extremes, which gives a crude but quite effective view of the skewness and spread of the data. This can be displayed in the following way:

```
# n (no. of observations)
M (depth of median)   median
F (depth of fourth)   lower fourth     upper fourth
  1                   lower extreme    upper extreme
```

so, for example, for the sodium ISFET data [37] used to produce the stem-and-leaf display of Figure 9, we have

```
#         98
M         49.5                129.0
F(or H)   25      123.25       134.0
          1       112.25       154.5
```

As already explained, the letters are used as tags or labels for the summary values. After the median and fourths, we have the eighths, sixteenths, thirty-seconds, and soon, denoted by the letters E, D, and C, etc. Table 10 shows the relationship of particular letter values to the area outside the limits they define for continuous distributions (the so-called tail area) [41].

Table 10 Letter Values and Their Relationship to the Tail Area[a] of Continuous Distributions

Letter value	Tag	Tail area
Median	M	0.5
Fourth or hinge	F or H	0.25
Eighth	E	0.125
Sixteenth	D	0.0625
Thirty-second	C	0.03125
Sixty-fourth	B	0.015625
	A	0.0078125
	Z	0.00390625
	X	0.001953125
	Y	0.0009765625

[a]Tail area = fraction remaining outside the letter value boundaries.

DATA ANALYSIS

So now we can make a more detailed letter value display for the sodium ISFET data as shown in Figure 11. This gives us a compact way of describing the data distribution and showing how the data are spreading into the tails.

The letter values have many other uses, some of which are described below [41]. For example, we can summarize spread using the *fourth spread*:

Fourth spread = upper fourth − lower fourth

or

$$d_F = F_U - F_L$$

This is very close to the interquartile range because fourths are nearly the same as the more traditional quartiles. The fourth spread may be used to calculate boundaries in the ordered data set that enable us to identify extreme values.

A useful boundary to set is 1.5 times the fourth spread beyond the upper and lower fourths; values beyond these limits are considered "outside" values [40]. For a Gaussian distribution, the fourths are at the mean ± 0.6745 (SD), and the outside cutoffs defined above are at ± 0.00349 (SD), so only 0.698% of a Gaussian data set lies outside these limits [41]. Heavy-tailed distributions will have more and light-tailed distributions fewer values outside these boundaries. It is very necessary to point out that in samples of data containing fewer than 500 observations, distinguishing outliers from heavy tailedness is not a particularly realistic exercise [41].

	DEPTH	LOWER	UPPER	MID	SPREAD
N=	98				
M	49.5		129.000	129.000	
F/H	25.0	123.250	134.000	128.625	10.750
E	13.0	119.500	140.500	130.000	21.000
D	7.0	119.000	144.250	131.625	25.250
C	4.0	116.500	146.000	131.250	29.500
B	2.5	114.750	147.875	131.313	33.125
A	1.5	113.375	152.000	132.688	38.625
	1	112.250	154.500	133.375	42.250

Figure 11 Detailed letter value display.

Box-and-Whisker Plots

These concepts can be used in graphical form as box plots (or box-and-whisker plots) [42,43], the simplest form of which is illustrated in Figure 12, and several data sets can be compared on one plot as in Figure 13. The concept can be extended to include additional boundaries as in Figure 14.

An important extension of these ideas is the *notched box plot* [44,45], in which a notch is placed in the box at the position of the median, the width of the notch being the confidence interval for the median. The interval usually chosen is for 95% confidence, although other values may be appropriate. It is calculated using the sign test procedure discussed earlier in the section on nonparametric statistics. The notched box plot is illustrated in Figure 15.

If two notched box-and-whisker plots for two different data sets are plotted together on the same display, we have the graphical equivalent of the Mann–Whitney–Wilcoxon unpaired comparisons test. If the two notches fail to overlap, then the difference between the medians is significant at the 95% level of confidence.

Extending this idea, plotting several notched box plots on the same display is rather like doing a graphical one-way analysis of variance [46]. Boxes

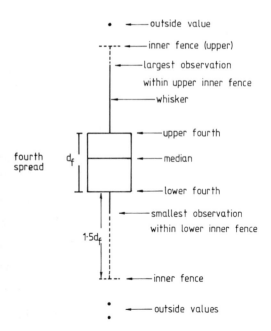

Figure 12 Simple skeletal box-and-whisker plot showing the principal features.

DATA ANALYSIS

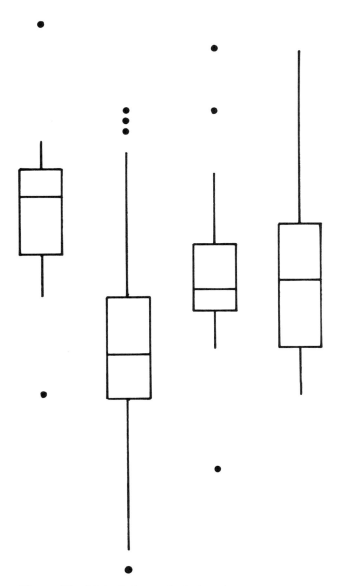

Figure 13 Multiple box-and-whisker plots.

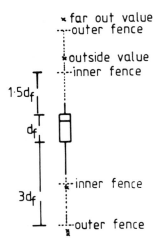

Figure 14 Extended version of the box-and-whisker plot showing additional boundaries.

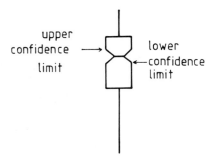

Figure 15 Notched box-and-whisker plot with the notches showing the 95% confidence limits on the median as calculated using the sign test.

whose notches overlap have medians that are not significantly different. This is illustrated in Figure 16. Care must be taken in the interpretation of such a multiple comparison, because the notches are not adjusted for the situation in which several hypotheses are being simultaneously tested [there $n(n-1)/2$ comparisons of pairs of medians if there are n data sets]. If the gap between any pair of notches is large, the conclusion is clearly against the validity of the null hypothesis. If the gap is fairly small, the conclusion is then less clear because, even under the hypothesis of equality of all the medians being compared, there

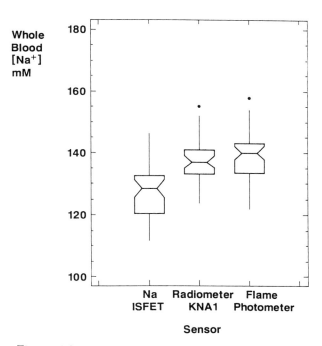

Figure 16 Multiple notched box-and-whisker plots: graphical one-way analysis of variance from observation of the overlap of the vertical positions of the notches.

is a 5% chance that one pair of notches will not overlap [46]. Under those circumstances, the multiple plot is a useful screening tool, but any conclusions should be confirmed by other tests or the collection of more observations.

Using Letter Values to Obtain Robust Estimates of Data Spread

We can obtain a resistant analog of standard deviation from the fourth spread from the knowledge that the F-spread of a Gaussian distribution is 1.349 × SD so the standard deviation should then be of the form

$$SD = \frac{\text{data F-spread}}{1.349}$$

This is known as the *F-pseudosigma*, and its square is the *F-pseudovariance* [41,47]. Other pseudosigmas and pseudovariances can be calculated from the other letter values.

If the F-pseudosigma and the standard deviation differ markedly, the former would be a better measure of spread, but we would also be alerted to hunting out the measurements that have increased the standard deviation. Using both measures of spread as a data screen would serve to eliminate faulty data arising from transcription or transmission errors or highlight the presence of outliers or a heavy-tailed data distribution.

Diagnosing Skewness and Elongation in Data Distributions

The letter values have another application in diagnosing skewness (departure from symmetry) and elongation (presence of heavy tails) [47]. This may enable us to distinguish, for example, a skewed or heavy-tailed distribution from a Gaussian distribution contaminated with outliers. If data are symmetrically distributed, the median will be the balance point, and the means of each pair of upper and lower letter values should be at the same point as well. These means are termed *midsummaries*, or *mids* for short [47]. If data are skewed to the right, as with the lognormal distribution, the mids should increase as we move outwards through the letter values toward the tails. Likewise, left skewness should produce decreasing mids. Figure 11 shows some right skewedness for the sodium ISFET blood [Na^+] values with increasing mids as we go down the table.

Elongation has to be defined by comparison with some suitable standard shape, such as the Gaussian [47]. The pseudosigma values, defined earlier, provide the basis for evaluating elongation. When the pseudosigmas associated with each letter value pair increase systematically, we diagnose heavy tails (more elongation than Gaussian); decreasing pseudosigmas indicate light tails. This had considerable advantages over using kurtosis because we can readily see whether one or two wild values might be the cause of the elongation.

Table 8 also lists pseudosigmas for sodium ISFET [Na^+] assays. The population of patients examined here gives us first an increase, then a decrease in the pseudosigmas. This suggests that we need to examine each tail separately because each tail may be behaving differently. We can do this using a *pushback technique* described by Hoaglin [47]. The processes involved are illustrated in Table 11.

First, we calculate the pseudosigmas corresponding to each pair of letter values. Then we find the median of these. This is then multiplied by the standard normal deviate corresponding to each of the lower and upper letter values. Those resulting numbers are then subtracted from the corresponding letter values to give *flattened letter values*. When these are plotted against the standard normal deviates (z), we should see a horizontal line if the data are normally distributed.

In Figure 17, we see the plot of the flattened letter values from Table 11,

DATA ANALYSIS

Table 11 Pseudosigmas, Flattened Pseudosigmas, and Standard Normal Deviates Corresponding to the Letter Values of [Na$^+$] Assays by a ISFET, as Listed in the Letter Value Display of Figure 8

Letter value tag	Pseudosigma	Lower flattened letter value	Upper flattened letter value	Standard normal deviate z for each tail	Square of standard normal deviate for both tails, z_p^2
F or H	7.9689	128.628	128.622	±0.674	1.8198
E	9.1265	128.676	131.324	±1.150	5.2946
D	8.2301	131.239	132.011	±1.534	9.4126
C	7.9195	131.364	131.136	±1.863	13.8756
B	7.6892	131.936	130.689	±2.154	18.567
A	7.9886	132.6676	132.7074	±2.418	23.377

Median value = 7.97875

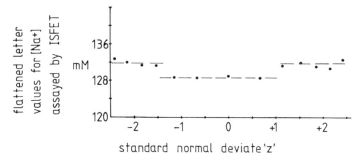

Figure 17 Disentangling the elongation components in each tail of a distribution by means of a plot of flattened letter values versus z, the standard normal deviate values. The plot uses data from assays, using a Thorn EMI ISFET, of whole blood [Na$^+$] from patients in critical care or undergoing cardiothoracic or liver transplant surgery.

which shows three horizontal zones, the two outer zones at about the same level. The tails seem to behave similarly but differ from the bulk of the data.

Plots of the mids versus z^2 and the pseudosigmas versus z^2 (Figures 18 and 19) provide for graphical tests for skewness and elongation patterns, respectively, both giving horizontal plots for Gaussian distributions. Clearly, there are deviations from normality, but because of the relatively small sample size (98

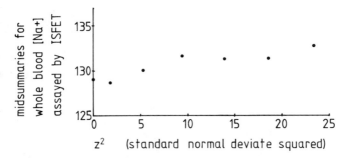

Figure 18 Testing for skewness using a plot of midsummaries versus z^2.

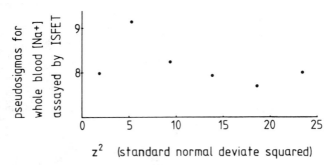

Figure 19 Testing for elongation using a plot of pseudosigmas versus z^2.

observations) it is probably wise not to be dogmatic about their nature. Hoaglin [47] discusses these shape evaluation techniques at much greater length, including their theoretical basis.

Resistant Regression

A useful exploratory regression method, which provides quite a considerable resistance against "wild" data, is Tukey's three-group resistant line [48,49]. This method has a breakdown bound of 1/6, so at most one-sixth of the data points can become arbitrarily wild without destroying the value of any regression line estimated. A computer algorithm for this method is described in detail by Velleman and Hoaglin [48], who give detailed FORTRAN and BASIC listings for the procedure. The FORTRAN version is incorporated into Minitab as RLINE. The procedure begins by sorting the (x,y) pairs into ascending order of increasing x; then the (x,y) pairs are divided into three groups, but the splitting of tied groups of x_i values is specifically avoided. Then we find three summary points, one for each group:

$$(x_L, y_L), (x_M, y_M), (x_R, y_R)$$

The x and y values in the summary pair are the medians of all the x and y values in that group. Initial estimates of the regression coefficient and intercept are

$$b_0 = \frac{y_R - y_L}{x_R - x_L}$$

and

$$a_0 = [(y_L - b_0\, x_L) + (y_M - b_0\, x_M) + (y_R - b_0\, x_M)]/3$$

Using these values, the residuals of the initial fit are calculated. These residuals are then fitted to the x variable in order to obtain an adjustment to the initially fitted regression. This process continues until the difference between successive values of the regression coefficient is within a specified limit.

Emerson and Hoaglin [49] give an updated review of the technique, in which they describe a method of overcoming the problem that sometimes the algorithm of Velleman and Hoaglin [48] fails to converge. Johnstone and Velleman [50] describe another algorithm for Tukey's resistant line that will always converge. Neither method seems to occur explicitly in Minitab, but, using BRIEF 4 with RLINE, one can obtain the slope at each iteration. Using this information as a starting point, one can then implement the simple, safe iteration procedure of Emerson and Hoaglin [49]. Johnstone and Velleman [50,51] used Monte Carlo simulations to test the reliability of Tukey's method and also of Theil's pairwise slopes method (see previous section), and both were found to behave favorably.

Emerson and Hoaglin [49] have described a very interesting approach to multiple regression using Tukey's three-group resistant line, in which one independent variable at a time is stripped out of the residuals remaining after the two-variable regression has been completed.

Identifying Unusual Observations

Turning now to the problem of determining quantitatively the influence of unusual observations in regression, the approach depends on whether we start from OLS regression or a robust, nonparametric, or exploratory viewpoint.

The conventional exploratory approach would involve examining diagnostic functions derived from the residuals obtained from an OLS regression. It is usual to consider two aspects: leverage and influence. Points of high leverage are from observations having an extreme value of one or more explanatory variables. In simple two-variable OLS regression we can define the following function for each of the n observations [49,52,53], often termed the "hat" matrix:

$$h_{ij} = \frac{1}{n} + \frac{(x_i - x)(x_j - x)}{\sum_{k=1}^{n}(x_k - x)^2}$$

For each point, h_{ii} (i.e, the h_{ij} in which $i=j$, the diagonal elements of the hat matrix) is regarded as measuring the leverage of that point (see Emerson and Hoaglin [49]). An observation with high leverage will grossly distort an OLS regression.

For two-variable problems, all the h_{ii} have values between $1/n$ and 1 and their sum is 2, so the average value is $2/n$. Constructing a stem-and-leaf display of the h_{ii} will enable us to identify values that are potentially excessive. (A point of high leverage does not necessarily appear as an outlier but may just be a long way from the midpoint of the regression.) One can also plot the leverage versus the index number of the data points.

The leverage can also be used to calculate a standardized residual, which is also useful for identifying unusual points:

$$r'_i = \frac{r_i}{s(1-h_i)^{1/2}}$$

These ideas can be readily extended to the multiple regression case.

Atkinson describes various methods of quantifying influence and leverage and their use in graphical exploratory analysis, principally for OLS regression diagnostics, and the interested reader is referred to his excellent monograph [53] for details. Another useful review of the exploratory approach to OLS residuals analysis is given by Goodall [52]. Carroll and Ruppert [54] review the disadvantages of using regression diagnostics based on the analysis of residuals from ordinary least squares. They suggest that a better strategy should involve the use of robust regression and an analysis of the residuals so obtained. One can then return to OLS regression, having identified the troublesome data points in a robust fashion. This is also the approach suggested by Hampel et al. [55], Atkinson [56], and Rousseeuw and Leroy [57].

With software now readily available for alternative approaches to regression (see end of Section 3.4 and bibliography), and with many authors now questionning whether one should begin with or focus solely on OLS methods, it would be sensible to compare OLS regression results with those from at least one alternative from among the families of exploratory, nonparametric, or robust regression methods. The comparison should include an examination of the diagnostic residuals plots from the alternative(s). Then one ought to have more

DATA ANALYSIS

confidence in whether the OLS approach is really appropriate. Robust regression diagnostics are discussed in Section 3.4.

Symbolic Exploratory Approaches to Multivariate Analysis

Our abilities to recognize similarities and differences between visual forms have led several researchers to propose the use of symbolic/graphical images for multivariate analysis. In some of these proposals, each variable measured for a particular specimen or case is coded by a specific feature of the symbol in a quantitative or semiquantitative manner. In other proposals, the variables are combined in some specific manner that affects the overall shape of the symbol. Four of the more important approaches are briefly discussed below. For further reading on this topic, see Seber [58], Krzanowski [59], Chambers et al. [60], du Toit et al. [61], and Thompson [62].

Chernoff Faces

In 1973, Chernoff [63] suggested that one could associate variables with different characteristics of the human face (e.g., eyes, ears, mouth, shape of head). The shape or size of this characteristic was related to the value of the variable so coded. It is often useful with this method to experiment with categorizing the different variables in order to obtain some simple stereotypes. We could choose correlate features of the face to reflect correlated variables (e.g., happy/sad mouth positions and presence/absence of frown lines could represent two correlated variables). Several groups have experimented with refinements of the original Chernoff face concept, including Flury and Riedwyl [64,65] and Schüpbach [66,67], who have produced software for use with a PC and plotter that enables one to code as many as 36 variables into an asymmetrical face.

Flury and Riedwyl [65] applied the asymmetrical faces software to the investigation of changes in certain population variables for the various Swiss cantons between the 1970 and 1980 censuses. Each canton is represented by a face in Figure 20. The right side of the face represents 1970 data, and the left represents 1980 data. For each side of the face, 11 variables are coded, and these can be grouped into five classes: foreign population, sex, age structure, religion, and marital status. From Figure 20 it is easy to recognize that cantons can be grouped in various ways at a glance. For example, the foreign population percentages of the various cantons are represented by the darkness of the hair. The curvature of the eyebrows represents the proportion of Protestants, and the eyebrow thickness the proportion of Catholics. So again, one can immediately identify clusters of cantons with a predominantly Catholic population.

In the same report [65], the technique was applied to a quality inspection of 79 tobacco lots and their comparison with the same measurements taken on an 80th lot after delivery. With this data set, nobody had considered that the 79

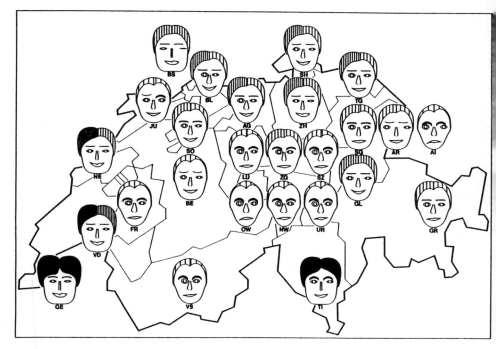

Figure 20 Asymmetrical Chernoff faces plotted on a map of Swiss cantons to illustrate the geographical distribution of various sociological characteristics of the populations of each canton. (Reproduced from reference 65, by kind permission of Professor Bernard Flury and Dr. Hans Riedwyl, Institut für Mathematische Statistik und Versicherungslehre, Universität Bern.)

standard lots could be clustered into 10 homogeneous subgroups until the faces were examined. Then it was observed that the faces fell into groups with an obvious resemblance to one another. The 80th lot had a face that had no resemblance to any of the groups. Thus Flury and Riedwyl suggest the potential use of asymmetrical Chernoff faces in quality control and to demonstrate differences between measured values (on one side of the face) and specification (on the other side).

With suitably optimized algorithms for transforming the data vector prior to mapping it to the face parameters, this could provide a potentially powerful tool for decision making about whether a process was in or out of control. It would also be desirable in such applications to optimize the assignment of

Andrews Curves

Andrews [68] proposed the use of finite Fourier series to represent multivariate data points and introduced the function

$$fX(t) = x_1 + x_2 \sin t + x_3 \cos t + x_4 \sin 2t + x_5 \cos 2t + \ldots$$

where the function is plotted over the range $-\pi < t < \pi$ and $x_1, \ldots, x_p$ are the components of the data vector X'. The curves have the following useful properties:

1. The function preserves means, so the plot for the mean vector resembles an "average" curve.
2. The function preserves Euclidean distances between any two vectors X' and Y', so points close together produce Andrews curves that are also close together.
3. The function preserves linear relationships, so that if Y' lies between X' and Z' then so do the corresponding curves.
4. The plot yields a one-dimensional projection of the data vector X' for any value of $t = t_0$ onto the vector

   $$a_0 = [1/(2)^{-1/2}, \sin t_0, \cos t_0, \sin 2t_0, \cos 2t_0, \ldots]$$

   and this allows us to detect both clusterings and outliers or other data peculiarities.
5. The function preserves variances, so that if the variables are uncorrelated in the data matrix with a common variance s^2, then the variance of the function is

   $$s_t^2 = s^2 (1/2 + \sin^2 t + \cos^2 t + \sin^2 2t + \cos^2 2t + \ldots)$$

However, because variances are frequently correlated among the components of the data vectors, Seber [58], Krzanowski [59], and du Toit et al. [61] suggest transforming X' to its vector of standardized principal components and associating these in order of importance with the order of the appearance of the frequencies in the series. Thus the first principal component would associate with x_1, the second with x_2, and so on.

An example of an Andrews curve plot is sketched in Figure 21.

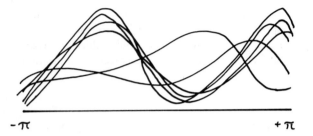

Figure 21 Sketch of a series of Andrews curves (Fourier-synthesized sine waves of multivariate data vectors).

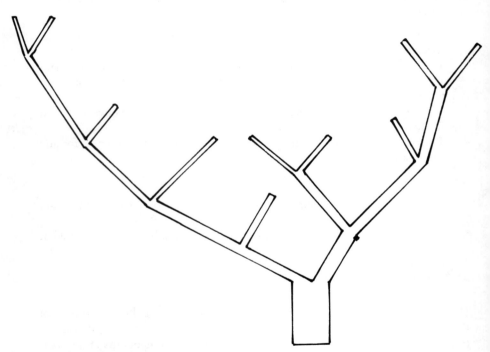

Figure 22 Kleiner–Hartigan tree plot.

Kleiner–Hartigan Trees

With many types of symbolic approaches to plotting multivariate data, the order of assigning variables to features of the symbol is very arbitrary. Replotting with different assignments may emphasize alternative features and relationships. Kleiner and Hartigan [69] proposed avoiding this arbitrariness by first performing a hierarchical clustering of the variables over all data vectors and then using the resulting dendogram as the basis for a *tree* symbol. A tree symbol is used for each data vector. Closely correlated variables are represented by branches close together on the same limb, and variables not so correlated are placed on different limbs. The length of each branch is determined by the value of the corresponding variable for that vector, and the length of a limb by the average of all the branches that it supports. A sketch of the tree concept is shown in Figure 22.

Star Plots

One can plot side by side the normalized values of variables in a data vector as a form of bar chart known as a *profile plot*. Welsch [70] suggested that instead of plotting the scalar values in a row, we plot each one as a line radiating from a common central point and join the far ends of the lines together to produce a polygonal shape termed a *star plot*. The assignment of variables to rays of the star is arbitrary, so it is in the interest of the user to try to optimize the assignment in some way, for example, by using cluster analysis (S.J. Haswell, personal communication). Figure 23 illustrates the display of star plots for a series of data vectors. More information on these plots can be found in Seber [58], Chambers et al. [71], Hamilton [72], and du Toit et al. [61].

Exploratory Methods for Two-Way Analysis of Variance

As discussed in Section 3.2, Friedman's two-way analysis of variance by ranks is unable to cope with missing values. Exploratory methods provide such a capability, which is particularly useful in observational work in such areas as clinical biochemistry and in the environmental sciences.

In a two-way table, the rows and columns represent two factors varying regularly from one end of the row or column to the other. The entries in each cell of the table represent the values of responses to the particular values of the row and column variables corresponding to that cell. The simplest form of model relating the response variable to the row and column variables is an additive model:

$$y_{ij} = m + a_i + b_j + e_{ij}$$

where y_{ij} is the response variable, m is a typical response over the whole table, a_i is the value of the row variable at level i, b_j is the value of the column variable

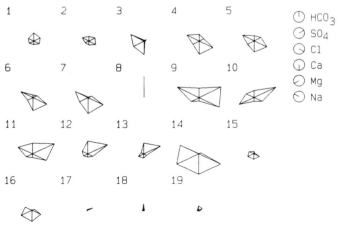

Figure 23 Star plot of the results of chemical analyses of 19 brine samples recovered from drill-stem tests of three carbonate rock units in Texas and Oklahoma. Original data from A.G. Ostroff, Comparison of some formation water classification systems, *Bull. Am. Assoc. Petroleum Geol.*, 51(3):404–416 (1967), cited in J.C. Davis, *Statistics and Data Analysis in Geology*, 2nd ed., John Wiley & Sons, Inc., New York, pp. 605–607. The plot was drawn using the Stata statistical software package (Computing Resource Center, Santa Monica, CA 90401).

at level j, and e_{ij} is the random departure from the expected additivity of the model at cell ij.

A fit of this type can be done iteratively by finding and subtracting row and column medians from the data table. So we could start by calculating the medians of the observations in each and every row. Then each median is subtracted from every observation in the corresponding row. The next step is to find the column medians and subtract each median from the observations in the corresponding column. If any row or column has a zero median, then obviously no change is made. The process can be continued until all the row and column medians a process known as *median polish* [73,74].

Velleman and Hoaglin [73] describe a computer program for fitting of two-way tables by median polish, the FORTRAN version of which is incorporated in Minitab. The method is resistant to outliers in the data table, which is certainly not true of least squares methods. It is often difficult to identify genuine outliers using least squares because the effect of the outliers may leak out from cells where they are located and *mask* the usual identification process using residuals [74]. This does not usually happen in median polish, in which outliers

DATA ANALYSIS

generally produce very obvious residuals. Systematic nonadditivity in the table can be identified by graphical exploration of the residuals [74].

Generalizations of median polish methods to cope with multiplicative and additive-plus-multiplicative fits are described by Emerson and Wong [75], and Cook [76] has described how the method can be extended to the analysis of three (or more) way tables.

Godfrey [77] has reviewed the use of the square combining table as an alternative resistant technique to median polish.

3.4 ROBUST METHODS

The nonparametric or distribution-free statistical methods described in Section 3.2 and the exploratory tools described in Section 3.3 are among the armory of tools that can be helpful as screening procedures when robust regression methods are being applied. For example, points of high influence or leverage can be identified and specifically downweighted in a weighted least squares regression, or they can be deleted in a stepwise fashion. The subject of robust regression is large and still rapidly developing, and no more than a sample of some of the more accessible developments can be given here.

This section will also introduce the reader to the topics of robust estimation of location (average value) and spread. In each case there is an opportunity for outlier detection, but that definitely does not mean outlier rejection. As R.A. Fisher [78] cautioned in 1922, "a point is never to be excluded on statistical grounds alone." Many outlier detection criteria work only under very specific conditions. It is quite possible that they may not detect *any* outliers if those conditions are violated (Hawkins [79]).

Outlier detection enables us to identify points for more detailed scrutiny, in order to determine whether transcription and/or transmission errors or other experimental artifacts have distorted such observations, or whether the observation(s) are genuine and from the tails of the underlying distribution. If the former is definitely the case, then we have sound criteria upon which to reject the observation(s). But if the latter is the case, then we might be looking at resistant or robust methods, such as those described in earlier sections, as a starting point for robust regression in which there is an attempt to model the underlying distribution or systematically account for it in a weighting scheme.

Beginning with location estimators, we have already seen that the mean is not resistant to the presence of wild values and is sensitive to all the values in a data set. The median is not sensitive to outliers and is sensitive only to the one or two middle values in the data set. Other possible estimators include the trimean, the broadened median (BMED), various trimmed means, the bisquare weight (known more generally as the biweight or Tukey's biweight), Andrews' wave estimator, Hampel's redescending estimators, and Huber's estimators, about which the reader may learn more elsewhere [80–83]. The choice of estimator to give the most reliable view of the average value of a data set continues to be an active area of research, but some simple guidelines are given by Rosenberger and Gasko [82]. For data sets containing n observations, they suggest the following.

1. For $n \leq 6$, use the median.
2. For $n = 7$, trim two observations from each tail of the ordered set and calculate the mean of the remaining observations.
3. For $n \geq 8$, trim 25% from each tail of the ordered set of observations and calculate the mean of the remaining observations. This is known as the midmean or 25% trimmed mean.

They also suggest that these rules seem to protect against outliers in both symmetric and asymmetric distributions and give reasonably good reliability in estimation. They are not the most optimum in all cases, but for many circumstances they will probably be quite adequate.

Of the more complicated estimators, Tukey's biweight appears very useful. Its calculation is by means of an iterative algorithm, and as a simplification the one-step version of the formula for this is

$$T_{bi} = M + \frac{\sum_{|u_i|<1} (x_i - M)(1 - u_i^2)^2}{\sum_{|u_i|<1} (1 - u_i^2)^2}$$

where M is the median, i is an index representing the observation number, $u_i = (x_i - M)/c$ MAD, c being a tuning constant (6 and 9 are values that have been demonstrated to work well), and MAD is the median absolute deviation from the median.

For robust estimations of scale (i.e., spread of the data) and of confidence intervals for location, several methods are available (see Iglewicz [83]). Starting with scale estimators, the simple ones include

1. The standard deviation (which is not robust to outliers)
2. The fourth spread (already discussed in the exploratory data analysis section), d_F
3. The absolute deviation from the sample median, AD
4. The median absolute deviation from the sample median, MAD

Evaluation of these with simulated data from a Gaussian distribution and two other symmetrical distributions, one being heavy-tailed and the other very heavy tailed, showed the fourth spread to be the best overall. The fourth spread still leaves much to be desired, but its great attraction is its computational simplicity. Of the more complicated estimators, ones based on Tukey's biweight appear most promising:

$$s_{bi} = \frac{n^{1/2} \left[\Sigma_{|u-i|<1} (x_i - M)^2 (1 - u_i^2)^4 \right]^{1/2}}{|\Sigma_{|u-i|<1} (1 - u_i^2) (1 - 5u_i^2)|}$$

where $u_i = (x_i - M)/c$ MAD, with the tuning constant $c = 9$.

From these ideas on robust location and scale estimation, we can go on to calculate robust confidence intervals based on the fourth spread and the median for simplicity of calculation, or on the biweight for greater robustness over a wider range of distribution types. The usual way of calculating a 100 $(1-\alpha)$% confidence interval for the mean is based on the use of Student's t:

$$x \pm ts/n^{1/2}$$

By analogy, the robust confidence interval can be constructed:

$$T \pm t^* w/n^{1/2}$$

where T and w are a pair of robust estimators of location and scale and t^* is a conservatively chosen constant analogous to the appropriate Student's t.

Thus, for hand computation, a robust confidence interval about the median is

$$M \pm \frac{t_{n-1}(d_F)}{1.075 n^{1/2}}$$

The more efficient robust interval based on biweights is

$$T_{bi} \pm t_{0.7(n-1)} \frac{s_{bi}}{n^{1/2}}$$

Appropriate values for t^* are (1) for the median-based formula, Student's t with $n-1$ degrees of freedom, and (2) for the biweight-based formula (for $n \geq 8$), Student's t for $0.7(n-1)$ degrees of freedom.

For hypothesis testing (e.g., the null hypothesis $\mu = \mu_0$ versus the alternative $\mu > \mu_0$), the biweight approach would be

$$\frac{T_{bi} - u_0}{s_{bi}/n^{1/2}} > t_{0.7(n-1)}$$

The Analytical Methods Committee of the Royal Society of Chemistry recently made proposals for the use of robust methods [84]. For location, they used the estimators H15 and A15, and for scale, a scaled version of the median absolute deviation of the median (divided by 0.6745, the distance of each quartile from the mean in a standardized Gaussian distribution) or a scaled median of the absolute deviations from the mean (again, the scaling was achieved by dividing by 0.6745). Their choices are not entirely supported in recent literature [80–83,85–87]. Diaconis [88] makes a case for the combined use of exploratory data analysis with other techniques of data analysis, to avoid the constraints of dogmatism in approaching an understanding of any given data set.

Tukey [42] gives an example of what can be missed by inadequate exploration of data using Rayleigh's measurements of the molecular weight of nitrogen prepared by chemically removing oxygen from air compared with the preparation of nitrogen by other methods including chemical decomposition of nitrous oxide, nitric oxide, or ammonium nitrite. If Rayleigh had not explored his data and noticed that the methods of preparing nitrogen gave molecular weights that fell into two distinct groups of measurements, he might not have discovered argon.

One should be neither dogmatic nor arbitrary in approaching data analysis, especially with small data sets that may or may not be typical of the sampled population. However, this also holds even for quite large data sets, as Mosteller and Tukey [89] point out.

Robust Regression Methods

This topic has been touched upon in earlier sections, with the nonparametric pairwise slopes method of Theil and the three-group resistant line of Tukey. Several additional methods devised by, respectively, Siegel [90], Rousseeuw [91], and Beaton and Tukey [92] are described briefly below.

Another Nonparametric Robust Regression Method

An extension of the pairwise slopes method, termed repeated median regression, was recently devised by Siegel [90] that can be implemented using the WSLOPES routine of Minitab as the starting point. The algorithm for the slope is

$$b_{RM} = \text{med}_{0 \text{ to } i} \{\text{med}_{j=i} \{b_{ij}\}\}$$

in other words, for each of the n points there are n-1 slopes that pass through that point, and the first stage is to take the medians of each of the n sets of $n-1$ slopes. Then we take the median of these n medians. There are n possible intercepts of the form

$$a_i = y_i - b_{RM}x_i$$

and the appropriate intercept for this regression is the median of these n possible intercepts:

$$a_{RM} = \text{med}_{0 \text{ to } i} \{a_i\}$$

This procedure has a breakdown bound of 1/2 and this is maximally protective against wild data. The major disadvantage of this method is the large compu-

DATA ANALYSIS

tational effort and RAM needed for even fairly moderate sizes of data sets on 8086-based PCs. This should not be a problem on 80386/80486 PCs or on workstations.

Rousseeuw's Least Median of Squares Regression

This method was published by Rousseeuw in 1984 [91] and, like Siegel's repeated median method, has a breakdown bound of 1/2. The approach is to minimize the median of the squared residuals instead of the sum. This method is described in great detail in the monograph of Rousseeuw and Leroy [57], which acts as the manual for software that is available for PC users direct from Rousseeuw (Prof. Peter J. Rousseeuw, Vesaliuslaan 24, B-2520 Edegem, Belgium). His software implements least squares, least median of squares (LMS), and reweighted least squares (RLS). RLS may be performed once outliers have been identified using LMS. It is possible to use LMS for multiple regression, time series analysis, and orthogonal regression. For those readers with workstations, Rousseeuw's LMS is incorporated into the package S-PLUS.

Robust Regression Using Tukey's Biweights

There are various ways of making OLS regression more robust, including weighted and iteratively weighted methods. In many statistical software packages available for use on personal computers, it is possible to do weighted least squares regression, but the choice of weighting factors is left to the user. This is obviously not helpful to the user without guidance on appropriate weighting schemes. However, useful approaches to iteratively weighted linear least squares, for either hand computation or simple implementation on a personal computer, have been described by Mosteller and Tukey [93]. The worked example below follows the scheme of Mosteller and Tukey for the use of biweights. For the alternative use of stepweights, which are useful in smaller calculations with a calculator, the reader is referred to Mosteller and Tukey [89].

The example given below uses the least squares regression results as the starting point.

Stage 1

Row	x	y	xy	x^2	
1	1.01	1.08	1.0908	1.0201	
2	1.98	2.02	3.9996	3.9204	
3	3.04	2.90	8.8160	9.2416	$\beta_1 = 1.07594$
4	4.07	4.89	19.9023	16.5649	
5	4.97	5.12	25.4464	24.7009	$n = 7$
6	6.02	7.50	45.1500	36.2404	
7	8.01	7.90	63.2790	64.1601	
Totals			167.684	155.848	

Stage 2

Fit$_1$ = $\beta_1 x$	$\|e\|$ = $\|y - \beta_1 x\|$	$\|u\|$ = $\|e\|/n^* \text{med}(\|e\|)$	w_1 = $(1-u^2)^2$	$w_1 xy$	$w_1 x^2$
1.08670	0.002582	0.002582	0.999987	1.0908	1.0201
2.13037	0.042514	0.042514	0.996388	3.9852	3.9191
3.27087	0.37087	0.142857	0.959600	8.4598	9.2016
4.37909	0.51091	0.196800	0.924040	19.6941	16.3916
5.34744	0.22744	0.087609	0.984708	25.4136	24.6690
6.47718	1.02282	0.393986	0.713645	43.3266	34.7768
8.61831	0.71831	0.276690	0.852747	62.3034	63.1709
			Totals	164.60	153.15

$\beta_2 = 1.06840$

$\|e\|$ is the absolute value of the residuals, and w_1 is first set of biweights.

Stage 3. This repeats the stages of calculation shown above but now using the stage 2 regression coefficient, β_2. The resulting new coefficient, β_3, is 1.07480; a further stage of calculation, using this coefficient as starting value, yields $\beta_4 = 1.06790$, and so on. The iteration of this calculation yields further coefficients, and the stopping criterion is determined by the data analyst setting an appropriate limit to the change between successive iterations.

The reader is invited to try this weighting scheme in the regression of the data in column 1 on the data in column 2 of Table 2. The biweight regression is implemented in the Solo statistical package from BMDP. Hamilton [94] also describes an implementation, as a macro in the student edition of the statistical software package Stata, of an iteratively reweighted least squares using Tukey's biweight function with a tuning constant that gives any residuals greater than ±7 times the median absolute value of the residuals a weight of zero. Using the Stata macro, one can plot the OLS and tuned Tukey's biweighted regression on the same graph for comparison. The same Stata macro command can also be used for robust multiple regression, and one can even graphically display the convergence process through the various stages of iteration. Stata, version 2.1 (released in 1990), now incorporates a modified biweight regression as well, with a weighting factor of 4.2.

An alternative to the use of OLS regression as the starting point would be a robust or resistant regression such as Tukey's three-group resistant line, Theil's method, or Rousseeuw's least median of squares method, described above. This can be followed by just one or a few iterations of Tukey's biweight method to give a highly efficient and robust procedure (see Gouying Li [95]). This approach has a breakdown bound of 1/2.

Robust Regression Diagnostics

A distinction was made between diagnostics for single outliers and those for multiple outliers, in Section 3.3 under the subhead Identifying Unusual Observations. Confusion can also arise from the inadequacies of many OLS regression diagnostics because of the possibility that "masking" of outliers can occur. In this respect, the use of OLS regression as a starting point for a reweighted least squares may not always provide the protection one desires. For example, Carroll and Ruppert [96] investigated the trimmed least squares estimator and found that it was not robust if the starting point was least squares. The use of robust diagnostics, in combination with robust/resistant regression methods, is a safer starting point.

The simplest robust regression diagnostics are the standardized residuals from a robust or resistant regression plotted as an index plot (i.e., the abscissas are the index numbers of the observations). This allows one to focus rapidly on potential outliers for further investigation.

Rousseeuw and Leroy [97] describe a resistant diagnostic, derived from an "outlyingness" parameter, that is an option in the output requested in Rousseeuw's software PROGRESS. Index plots of this diagnostic are a useful aid in identifying outliers. Giltinan et al. [98] have introduced some new estimation methods for weighted regression that are useful for cases where the variance is not homogeneous (this property is termed heteroscedasticity), the FORTRAN code for which is available from Giltinan. An example of a specific kind of heteroscedasticity is the case where the variance is proportional to the magnitude of the variable. Problems of heteroscedasticity occur frequently in such areas as chemical/enzyme kinetics and immunoassay. Carroll and Ruppert [99] have described a useful robust diagnostic with good outlier discrimination powers, but as yet there is no readily available specific software that would allow its use to become widespread. They did their calculations with the GAUSS matrix programming language or with APL through the STATGRAPHICS package.

REFERENCES

1. W.J. Conover, *Practical Nonparametric Statistics*, 2nd ed., Wiley, New York, Introduction, pp. 1–14. (1980). (Other references to this book use the abbreviation PNS.)
2. F.R. Hampel, E. Ronchetti, P.J. Rousseeuw, and W.A. Stahel, *Robust Statistics—*

To save space, abbreviations are used in citing portions of Works for which full information has been supplied in an earlier listing. These abbreviations and corresponding primary references are: **ABCs** *of EDA* [45], ANSM [15], DAR [3], EDA [4], EDTTS [47], GMDA [46], NMQA [14], PNS [1], RRDD [57], RS [2], UREDA [5].

The Approach Based on Influence Functions, Wiley, New York, Chapter 1. (1980). (Other references to this book use the abbreviation *RS*.)

3. F. Mosteller and J.W. Tukey, *Data Analysis and Regression—A Second Course in Statistics*, Addison-Wesley, Reading, Mass., Chapter 1. (1977). (Other references to this book use the abbreviation *DAR*.)
4. J.W. Tukey, *Exploratory Data Analysis*. Addison-Wesley, Reading, Mass. (1977). (Other references to this book use the abbreviation *EDA*.)
5. D.C. Hoaglin, F. Mosteller, and J.W. Tukey, eds. *Understanding Robust and Exploratory Data Analysis*, Wiley, New York, Introduction. (1983). (Other references to this book use the abbreviation *UREDA*.)
6. J.M. Thompson, S.C.H. Smith, R. Cramb, and P. Hutton, Evaluation of ion-selective field effect transistors for measurement of ions in whole blood, *Methodology and Clinical Applications of Ion-Selective Electrodes*, Vol. 10, International Federation of Clinical Chemistry, IFCC Technical Secretariat, Copenhagen, Denmark, pp. 305–313. (1989).
7. W.E. Deming, *Statistical Adjustment of Data*, Wiley, New York (1943); reprinted by Dover, New York (1964).
8. F.R. Hampel, A general qualitative definition of robustness, *Ann. Math. Stat.*, 42:1887–1896 (1971).
9. J.M. Thompson, S.H.C. Smith, R. Cramb, P. Hutton, J.P. Millns, and R. Ward, Clinical evaluation of sodium ion selective field effect transistors, *Analyst* (1991), in press.
10. P.J.M. Wakkers, H.B.A. Hellendoorn, G.J. Op de Weegh, and W. Heerspink, Applications of statistics in clinical chemistry. A critical evaluation of regression lines, *Clin. Chim. Acta.*, 64:173–184 (1975).
11. P.J. Cornbleet and N. Gochman, Incorrect least-squares regression coefficients in method-comparison analysis, *Clin. Chem.*, 25:432–438 (1979).
12. J. Mandel, *The Statistical Analysis of Experimental Data*, Wiley, New York (1964); reprinted by Dover, New York (1984), pp. 288–292.
13. W.J. Conover, *PNS*, pp. 105–116.
14. J.D. Gibbons, *Nonparametric Methods for Quantitative Analysis*, Holt, Rinehart and Winston, New York, pp. 94–122. (1976). (Other references to this book use the abbreviation *NMQA*.)
15. P. Sprent, *Applied Nonparametric Statistical Methods*, Chapman and Hall, London, pp. 36–42. (1989). (Other references to this book use the abbreviation *ANSM*.)
16. Minitab, Inc., *Minitab Reference Manual—Release 6.1*, Minitab, Inc., State College, Pa., pp. 191–193. (1988).
17. W.J. Conover, *PNS*, pp. 278–292.
18. J.D. Gibbons, *NMQA*, pp. 406–408.
19. P. Sprent, *ANSM*, p. 236.
20. W.J. Conover, *PNS*, pp. 215–227, 448–452.
21. J.D. Gibbons, *NMQA*, pp. 159–173, 409–416.
22. P. Sprent, *ANSM*, pp. 87–99.
23. W.J. Conover, *PNS*, pp. 229–237.
24. J.D. Gibbons, *NMQA*, pp. 174–193.

25. P. Sprent, *ANSM*, pp. 112–115, 119–122.
26. W.J. Conover, *PNS*, pp. 239–248.
27. P. Sprent, *ANSM*, pp. 102–103.
28. W.J. Conover, *PNS*, Chapter 6.
29. J.D. Gibbons, *NMQA*, pp. 56–77, 250–258.
30. P. Sprent, *ANSM*, pp. 48–57, 104–106.
31. P. Sprent, *ANSM*, pp. 152–155.
32. J.S. Maritz, *Distribution-Free Statistical Methods* Chapman and Hall, London, pp. 161–170. (1981).
33. J.M. Thompson, The use of a robust and resistant regression method for personal monitor validation with decay of trapped materials during storage, *Anal. Chim. Acta.*, 186:205–212 (1986).
34. J.F. Lancaster and D. Quade, A nonparametric test for linear regression based on combining Kendall's tau with the sign test, *J. Am. Statist. Assoc.*, 80:393–397 (1985).
35. W.J. Conover, *PNS*, pp. 299–308.
36. P. Sprent, *ANSM*, pp. 122–126.
37. J.M. Thompson, unpublished results.
38. Mark Twain, *Autobiography*.
39. J.D. Emerson and D.C. Hoaglin, Stem and leaf displays, *UREDA*, Chapter 1.
40. F. Mosteller and J.W. Tukey, *DAR*, pp. 45–48.
41. D.C. Hoaglin, Letter values: a set of selected order statistics, *UREDA*, Chapter 2.
42. J.W. Tukey, *EDA*, pp. 39–56.
43. J.D. Emerson and J. Strenio, Boxplots and batch comparison, *UREDA*, Chapter 3.
44. R. McGill, J.W. Tukey, and W.A. Larsen, Variations of boxplots, *Am. Statist.*, 32:12–16 (1978).
45. P.F. Velleman and D.C. Hoaglin, *Applications, Basics and Computing of Exploratory Data Analysis*, Duxbury Press, Boston, Mass., Chapter 3. (1981). (Other references to this book use the abbreviation *ABCs of EDA*.)
46. J.M. Chambers, W.S. Cleveland, B. Kleiner, and P.A. Tukey, *Graphical Methods for Data Analysis*, Wadsworth International Group, Belmont, Calif., pp. 60–63. (1987). (Other references to this book use the abbreviation *GMDA*.)
47. D.C. Hoaglin, Using quantiles to study shape, in *Exploring Data Tables, Trends and Shapes* (D.C. Hoaglin, F. Mosteller, and J.W. Tukey, eds.), Wiley, New York, Chapter 10. (1985). (Other references to this book use the abbreviation *EDTTS*.)
48. P.F. Velleman and D.C. Hoaglin, *ABCs of EDA*, Chapter 5.
49. J.D. Emerson and D.C. Hoaglin, *UREDA*, Chapter 5.
50. I.M. Johnstone and P.F. Velleman, Tukey's resistant line and related methods: asymptotics and algorithms, Proceedings of the Statistical Computing Section, American Statistical Association, pp. 218–223. (1981).
51. I.M. Johnstone and P.F. Velleman, The resistant line and related regression methods, J. Am. Statist. Assoc., 80:1041–1054 (1985). (Including the comments by J.W. Tukey in the same journal issue, pp. 1055–1059.)
52. C. Goodall, Examining residuals, *UREDA*, Chapter 7.
53. A.C. Atkinson, *Plots, Transformations and Regression—An Introduction to Graphical*

Methods of Diagnostic Regression Analysis, Oxford Univ. Press, Oxford, England, Chapter 2. (1985).
54. R.J. Carroll and D. Ruppert, Transformation and Weighting in Regression, Chapman and Hall, London, Introduction. (1988).
55. F.R. Hampel, E.M. Ronchetti, P.J. Rousseeuw, and W.A. Stahel, RS, pp. 330–331.
56. A.C. Atkinson, Masking unmasked, Biometrika, 73:533–541 (1986).
57. P.J. Rousseeuw and A.M. Leroy, Robust Regression and Outlier Detection, Wiley, New York, Introduction and Chapter 6. (1987). (Other references to this book use the abbreviation RROD.)
58. G.A.F. Seber, Multivariate Observations, Wiley, New York, Chapter 4. (1984).
59. W.J. Krzanowski, Principles of Multivariate Analysis—A User's Perspective, Oxford Univ. Press, Oxford, England, Chapter 2. (1988).
60. J.M. Chambers, W.S. Cleveland, B. Kleiner, and P.A. Tukey, GMDA, Chapter 5.
61. S.H.C. du Toit, A.G.W. Steyn, and R.H. Stumpf, Graphical Exploratory Data Analysis, Springer-Verlag, New York, Chapter 4. (1986).
62. J.M. Thompson, Visual representation of data including graphical exploratory data analysis, in Methods of Environmental Data Analysis (C.N. Hewitt, ed.), Elsevier Applied Science, Barking, Essex, England, Chapter 8. (1991).
63. H. Chernoff, The use of faces to represent points in k-dimensional space graphically, J. Am. Statist. Assoc., 68:361–368 (1973).
64. B. Flury, Construction of an asymmetrical face to represent multivariate data graphically, Technische Bericht, No. 3, Institut für Mathematische Statistik und Versicherungslehre, Universität Bern, Switzerland. (1980).
65. B. Flury and H. Riedwyl, Some applications of asymmetrical faces, Technische Bericht No. 11, Institut für Mathematische Statistik und Versichungslehre, Universität Bern, Switzerland. (1983).
66. M. Schüpbach, ASYMFACE: asymmetrical faces on IBM and Olivetti PC, Technische Bericht No. 16, Institut für Mathematische Statistik und Versicherungslehre, Univesität Bern, Switzerland. (1984).
67. M. Schüpbach, ASYMFACE: asymmetrical faces in TurboPascal, Technische Bericht No. 25, Institut für Mathematische Statistik und Versicherungslehre, Universität Bern, Switzerland. (1987).
68. D.F. Andrews, Plots of high-dimensional data, Biometrics, 28:125–136 (1972).
69. B. Kleiner and J.A. Hartigan, Representing points in many dimensions by trees and castles (with discussion), J. Am. Statist. Assoc., 76:260–276 (1981).
70. R.E. Welsch, Graphics for data analysis, Comput. Graphics, 2:31–37 (1976).
71. Chambers et al., GMDA, pp. 159–162.
72. L.C. Hamilton, Statistics with STATA, Brooks/Cole, Pacific Grove, Calif., pp. 69–70. (1990).
73. P.F. Velleman and D.C. Hoaglin, Median polish, ABCs of EDA, Chapter 8.
74. J.D. Emerson and D.C. Hoaglin, Analysis of two-way tables by medians, UREDA, Chapter 6.

75. J.D. Emerson and G.Y. Wong, Resistant nonadditive fits for two-way tables, *EDTTS*, Chapter 3.
76. N.R. Cook, Three-way analyses, *EDTTS*, Chapter 4.
77. K. Godfrey, Fitting by organised comparisons: the square combining table, *EDTTS*, Chapter 2.
78. R.A. Fisher, On the mathematical foundations of theoretical statistics, *Phil. Trans. Roy. Soc.*, 222A:322 (1922).
79. D.M. Hawkins, *Identification of Outliers*, Chapman and Hall, London. (1980).
80. D.C. Hoaglin, F. Mosteller, and J.W. Tukey, Introduction to more refined estimators, *UREDA*, Chapter 9.
81. C. Goodall, M-Estimators of location: an outline of the theory, *UREDA*, Chapter 11.
82. J.L. Rosenberger and M. Gasko, Comparing location estimators: trimmed means, medians and trimean, *UREDA*, Chapter 10.
83. B. Iglewicz, Robust scale estimators and confidence intervals for location, *UREDA*, Chapter 12.
84. Analytical Methods Committee of the Royal Society of Chemistry, Robust statistics—how not to reject outliers. Part 1. Basic concepts, *Analyst*, 114:1693–1697 (1989).
85. F.R. Hampel et al., *RS*, Chapter 2.
86. D.C. Hoaglin, Using quantiles to study shape, *EDTTS*, Chapter 10.
87. D.C. Hoaglin, Summarizing shape numerically: the g-and-h distributions, *EDTTS*, Chapter 11.
88. P. Diaconis, Theories of data analysis: from magical thinking through classical statistics, *EDTTS*, Chapter 1.
89. F. Mosteller and J.W. Tukey, *DAR*, Chapters 1 and 2.
90. A.F. Siegel, Robust regression using repeated medians, *Biometrika*, 69:242–244 (1982).
91. P.J. Rousseeuw, Least median of squares regression, *J. Am. Statist. Assoc.*, 79: 871–880 (1984).
92. A.E. Beaton and J.W. Tukey, The fitting of power series, meaning polynomials, illustrated by band-spectroscopic data, *Technometrics*, 16:147–185 (1974).
93. F. Mosteller and J.W. Tukey, *DAR*, pp. 356–361.
94. L.C. Hamilton, *Statistics with STATA*, Brooks/Cole, Pacific Grove, Calif., pp. 149–153. (1990).
95. Gouying Li, Robust regression, *EDTTS*, Chapter 8.
96. R.J. Carroll and D. Ruppert, *Transformation and Weighting in Regression*, Chapman and Hall, London, Chapter 6. (1988).
97. P.J. Rousseeuw and A.M. Leroy, *RROD*, pp. 237–245.
98. D.M. Giltinan, R.J. Carroll, and D. Ruppert, Some new estimation methods for weighted regression when there are possible outliers, *Technometrics*, 28:219–230 (1986).
99. R.J. Carroll and D. Ruppert, Diagnostics and robust estimation when transforming the regression model and the response, *Technometrics*, 29:287–290 (1987).

SOFTWARE AND BIBLIOGRAPHY

Statistical Software Packages for Personal Computers

The packages listed below vary in the degree of sophistication required of the user, in the degree of comprehensiveness of routines available, and in the flexibility with which new user-specific routines can be programmed. The list is not intended to be comprehensive, nor in any order of merit.

Minitab (supplied in the U.S.A. by Minitab, Inc., State College, PA 16801 and in the U.K. by Cle.Com. Ltd., Birmingham B15 2SQ). A well-known package used frequently for teaching, Minitab is command-driven (written in FORTRAN), and one can write specific new procedures as programs in the command language. This flexibility makes it a very useful tool for the types of data analysis discussed in this chapter. The graphics produced on a dot matrix printer are more primitive than can be achieved with a plotter. Exploratory and nonparametric routines are standard features.

Statgraphics (supplied in the U.S.A. by STSC Inc., Rockville, MD 20852 and in the U.K. by Mercia Software Ltd., Birmingham B7 4BJ). This powerful package is written in APL, and although the package is menu driven, the user may add in extra programs written in APL. The graphics available are very good even on dot matrix printers. The menu of statistical routines available is very wide-ranging including many exploratory and nonparametric procedures.

Stata (supplied by Computing Resource Center, Los Angeles, CA 90064). A relatively new and powerful package (originally launched in 1985), written in C, and having both command- and menu-driven modes and easy switching to and fro between them. It is possible to use Stata in various ways: like a statistical package, with a wide variety of procedures available as commands or as kits of programs; like a spreadsheet program, for what-if analyses; like a database management system; like a programming language to design new macro commands and complete programs; and like a graphics package, with a powerful graphics editor [enabling one to produce high-quality, high-resolution graphics even on dot matrix printers, with a WYSIWYG (what you see is what you get) capability]. Available for both MS DOS and Unix.

CSS Statistica (supplied in the U.S.A. by Statsoft, Inc., Tulsa, OK 74104 and in the U.K. by Eurostat Ltd., Letchworth SG6 3DA) is a very large and powerful integrated statistical package with database management, high-quality graphics, a wide range of statistical procedures, and expandability via a development environment that allows CSS to be used as an "open architecture" system supporting QuickBASIC, Pascal, C, and FORTRAN. This latter feature enables one to design new data analysis procedures, develop customized data acquisition/communication systems, and perform complex data transformation on CSS data files. The main package is run in menu mode. It offers a useful range of nonparametric and exploratory methods.

Systat (supplied in the U.S.A. by Systat, Inc., Evanston, IL 60201 and in the U.K. by Eurostat, the CSS suppliers) is a very powerful statistics and graphics package available in versions for MS DOS, Unix, Vax, and Macintosh. It operates in command mode and offers extensive capabilities in database management, high-resolution graphics via the SYGRAPH package, and a comprehensive range of statistical procedures, and as a programming language to extend the capabilities to custom-designed statistical/mathematical procedures.

Solo (supplied by BMDP Statistical Software, Inc., Los Angeles, CA 90025 and Cork, Eire) is a menu-driven package with a wide range of procedures, including many exploratory, robust, and nonparametric methods, database and spreadsheet, and high-resolution graphics, including various dynamic features such as real-time rotation of three-dimensional plots and highlighting of individual cases and groups.

BMDP (supplied as above for Solo) was originally designed as a comprehensive mainframe package and is now available as a PC version called PC90.

SPSS/PC+ (supplied by SPSS Inc., Chicago, IL 60611 and SPSS UK Ltd., Walton on Thames, Surrey KT12 5LU) is the PC version of the well-known mainframe package.

PC-ISP/DGS (supplied by Artemis Systems, Inc., Carlisle, MA 01741 and Datavision AG, 7250 Klosters, Switzerland) is a powerful statistical programming language with dynamic graphics capabilities and comes with a collection of macros. There is a users group with a macro exchange scheme.

Press et al. have published algorithms for least squares and least absolute deviations regression as FORTRAN, C, and Pascal listings and as floppy disks containing those and many other useful programs and worked examples as part of a large collection of numerical methods in the book *Numerical Recipes—The Art of Scientific Computing* (Cambridge University Press, from whom the floppy disks may also be purchased).

Bibliography of Useful Texts Not Listed in the References

J.C. Davis, *Statistics and Data Analysis in Geology*, 2nd ed., Wiley, New York. (1986). This book deals with nonparametric statistics and exploratory techniques such as cluster analysis as well as parametric techniques and comes with a floppy disk containing statistical software STAT, which performs some of the procedures contained in the book. STAT will run on any IBM-compatible PC with MS DOS (version 2.10 or higher) and 128K or more of memory. A more extensive library of programs known as TERRASTAT, for almost all of the procedures in the book may be purchased from Terrasciences, Inc., 7555 West Tenth Avenue, Lakewood, CO 80215.

N.M.S. Rock, *Numerical Geology—A Source Guide, Glossary and Selective Bibliography to Geological Uses of Computers and Statistics*, Springer-Verlag, Berlin, Germany. (1988). A very detailed source of references to a wide range of mathematical and statistical techniques, including robust, nonparametric, and exploratory methods and brief critical discussion on their advantages and disadvantages.

R.A. Thistead, *Elements of Statistical Computing—Numerical Computation*, Chapman and Hall, London, England. (1988). A useful discussion of the current state of the art and an excellent source of ideas for new areas of research.

J. Bertin, *Semiology of Graphics—Diagrams, Networks and Maps* (translated by W.J. Berg), University of Wisconsin Press, Madison, WI 53715. (1983). A very important text on graphical presentation of data.

C.F. Schmid, *Statistical Graphics*, Wiley, New York, NY. (1983). Another very important text in this area.

4
Calibration

John H. Kalivas *Idaho State University, Pocatello, Idaho*

In order for an instrumental response to be transformed into a more informative chemical variable such as concentration, it is necessary to calibrate the instrument. Ideally, the calibration should be designed to obtain accurate and precise estimates of sample concentrations. In this context, sampling theory explores various approaches available to ensure that acceptable concentration estimates are obtained from calibration models. Primary emphasis is on proper calculations for determining the appropriate number of samples needed for a chemical analysis such that the average value reported equals the true value within a given confidence interval.

Section 4.2, Least Squares and Linear Calibration, provides a basis for adequate understanding of how calibration parameters relating instrumental responses to concentrations can be obtained. The emphasis is on univariate regression because routine analyses performed in laboratories are usually concerned with single-component determinations. The discussion is extended to multivariate regression for multicomponent determinations, and there is some mention of biased regression methods.

Section 4.3 discusses confidence intervals, emphasizing univariate calibration with examples. This discussion is also extended to multicomponent analysis.

Occasionally, responses obtained from an instrument are not linearly proportional to concentration and require nonlinear calibration. Chapter 5 develops nonlinear regression comprehensively. Therefore, this chapter contains only a brief overview.

A moderate amount of confusion over detection limits prevails in the literature because of the variety of computational methods and terminology in the field. The principal disagreement rests in the interpretation of "statistically different" for a blank and sample. This chapter will try to clear the air and concentrate on univariate calculations. Both univariate and multivariate approaches for quantitating sensitivities of calibrations are developed in Section 4.5.

Of essential importance for all calibration procedures is the proper treatment of interferences. Section 4.6 concentrates on calibration methods that correct for interferences that are spectral in origin. Section 4.7 introduces the standard addition method, an effective calibration approach that corrects for chemical and physical interference effects.

The purpose of this chapter is not to discuss in detail the mathematics and statistics required to cover all aspects of calibration. Only details that are necessary to provide a basis for adequate comprehension are introduced. Examples are given that illustrate fundamental concepts to give the reader the necessary understanding to grasp more mathematically sophisticated techniques, which are amply referenced. Most of the described computations were performed using commercially available software for personal computers.

4.1 SAMPLING THEORY

Sampling is based on the fundamental principle that each sample is representative of the total material (population) from which it is taken. All units present in the population should have an equal probability of existing in the representative sample. For example, all chemical species (units) present in the body of a lake (population) should have equal chances of appearing in a sample of lake water. Similarly, if subsamples are to be taken from one sample of the lake, all chemical species should have equal probabilities.

Unfortunately, samples taken from nonhomogeneous populations do not contain units that exist with equal probabilities. Hence, an inherent sampling error always transpires owing to the heterogeneity of the population. In addition, an analysis error exists that is caused by random error present in the measurements used for the analysis. The resulting final variance, s^2, can be represented as

$$s^2 = s_s^2 + s_a^2 \qquad (1)$$

where s_s^2 denotes the sampling variance and s_a^2 represents the variance due to analysis. The discussion that follows pertains primarily to the sampling of homogeneous populations; there is some mention of sampling heterogeneous populations near the end of this section.

Normal Distribution

Consider the situation in which a chemist samples a lake to dryness taking N samples of equal convenient sizes. Chemical analyses are performed on each sample for concentration determinations of various chemical species. If normal distributions are followed, the probability function curves for each species should appear as shown in Figure 1, where μ specifies the population mean concentration for a species and x represents an individual concentration value for that species. The probability function for a normal distribution is given by

$$f(x) = \frac{1}{\sigma\sqrt{2\pi}} \exp\left[-\frac{(x-\mu)^2}{2\sigma^2}\right]$$

where σ is the standard deviation for this normal distribution. The standard deviation is calculated by using the equation

$$\sigma = \left[\frac{\sum_{i=1}^{i=N}(x_i-\mu)^2}{N}\right]^{1/2} \qquad (2)$$

where N expresses the number of samples needed to drain the lake. Appropriate sample sizes, such as volume or mass, are discussed later. The normal concentration distribution can be described as $x = N(\mu,\sigma^2)$, where σ^2 is termed the *variance*. Since the total population was sampled, μ and σ are the population parameters. Clearly, sampling the lake to dryness is not reasonable, and another, more feasible, approach is to use an assembly of n samples. In this case, $\bar{x}$ (the mean of the n samples taken) estimates μ, and σ is estimated by s (the standard deviation for the n samples), which is calculated as

$$s = \left[\frac{\sum_{i=1}^{i=n}(x_i-\bar{x})^2}{n-1}\right]^{1/2} \qquad (3)$$

and the normal concentration distribution is now represented as $x = N(\bar{x},s^2)$.

Standard Normal Distribution

For convenience, a normal distribution may be transformed to a standard normal distribution where the mean is zero and the standard deviation equals 1. Transformation occurs by letting

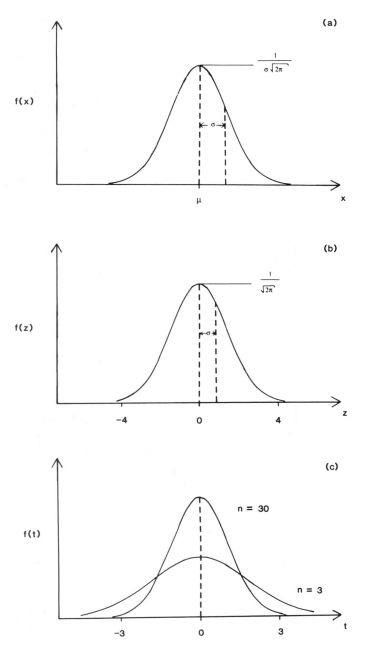

Figure 1 Distribution curves: (a) Normal, (b) standard normal, and (c) t distribution.

CALIBRATION

$$z_i = (x_i - \mu)/\sigma$$

The probability distribution can now be expressed as

$$f(z) = \frac{1}{\sqrt{2\pi}} \exp\left[-\frac{z^2}{2}\right]$$

represented by $z = N(0, 1)$. Figure 1 shows a plot for the standard normal distribution. In terms of our lake example, the normal concentration distributions for each chemical species with their different means and standard deviations can be transformed to $z = N(0, 1)$. A single table of probabilities, which can be found in most standard statistical books, can now be used.

Central Limit Theorem

According to the important theorem known as the *central limit theorem*, if n values are obtained from a non-normal distribution, the probability distribution for the mean will approach the normal distribution as n becomes large. For example, as more samples are selected from a lake with some concentration population mean and standard deviation, the concentration distribution of $\bar{x}$ for the samples will tend toward a normal distribution with mean μ and standard deviation $\sigma_{\bar{x}} = \sigma/\sqrt{n}$. Hence, a concentration mean of n samples is a random sample itself symbolized by $N(\mu, \sigma_{\bar{x}})$.

Implications of the Central Limit Theorem

With the central limit theorem, we have expanded from dealing with individual concentration determinations to concentration means. Each mean chemical species distribution can be transformed to a standardized distribution by

$$z = (\bar{x} - \mu)/\sigma_{\bar{x}}$$

with $\bar{x}$ functioning as the random variable. Provided that σ is known, the population mean can now be estimated to lie in the range

$$\mu = \bar{x} \pm z\,\sigma_{\bar{x}} \tag{4}$$

where z is obtained for the desired level of confidence from a table of probabilities.

t Distribution

When σ is not known, s must be used, requiring a new relationship to replace Eq. (4). The t distribution shown in Figure 1, for various numbers of samples, replaces the distribution based on z values. When $n \geq 30$, the t distribution approaches the standardized normal distribution. The interval in which the population mean will lie can be estimated using the equation

$$\mu = \bar{x} \pm ts_{\bar{x}} \tag{5}$$

where t expresses a value for $n-1$ degrees of freedom at the desired confidence level. The term *degrees of freedom* refers to the number of independent deviations $(x_i - \bar{x})$ that are used in calculating s. Equation (5) does not imply that the sample means are not normally distributed, but suggests that s is a poor estimate of σ except when n is large.

Other Distributions

Other statistical distributions exist in addition to those mentioned. The binomial, chi-square (χ^2), and F distributions give important statistical information. References 1 and 2 discuss these distributions further.

Hypothesis Testing

Hypothesis testing can be used to determine if a significant difference between μ and $\bar{x}$ exists and how many samples are required for the determination. For example, a person may question whether the concentration determined for a certified material significantly differs from the accepted value. As an illustration, you wish to use a standard copper solution labeled 50 ppm for a calibration, and you need to be sure the solution concentration is indeed 50 ppm. The hypothesis is then $\mu = 50$. To test this hypothesis you need to decide, among other factors elaborated below, how many samples should be taken from the standard solution and analyzed to ascertain the truth of the label. Hypothesis testing has many other uses, including testing variances, proportions, or the difference of two means. Here, we will emphasize determining the proper number of samples.

Null and Alternative Hypotheses and Types of Errors

A statistical hypothesis denotes a statement about one or more parameters of a population distribution requiring verification. The *null hypothesis*, H_0, designates the hypothesis being tested. If the tested H_0 is rejected, the *alternative hypothesis*, H_1, must be accepted. Using the mean as an example, the hypotheses can be expressed in the following forms:

$$H_0: \mu \geq \mu_0 \quad H_1: \mu < \mu_0$$
$$H_0: \mu \leq \mu_0 \quad H_1: \mu > \mu_0$$
$$H_0: \mu = \mu_0 \quad H_1: \mu \neq \mu_0$$

where μ symbolizes the sampling distribution mean and μ_0 signifies the hypothesized population mean. Two statistical tests are commonly used in hypothesis testing of means. When the standard deviation is known for the sample distribution, the z statistical test can be used.

CALIBRATION

$$z = (\bar{x} - \mu_0)/\sigma_{\bar{x}}$$

If the standard deviation is estimated as s, the t statistical test is used.

$$t = (\bar{x} - \mu_0)/s_{\bar{x}}$$

The appropriate decision in then made by comparing the calculated test statistic to theoretical values tabulated for respective levels of significance and degrees of freedom. These tests are covered in Chapter 2.

When testing the null hypothesis, acceptance or rejection errors are possible. Rejecting the null hypothesis when it is actually true commits a type I error. Likewise, accepting the null hypothesis when it is false results in a type II error. The chemist fixes the probability of making a type I error by specifying the level of confidence (or significance), α. If $\alpha = 0.05$, the probability of making a type I error translates to 0.05 (5%) and the probability for correct acceptance of H_0 becomes $1 - \alpha$, or 0.95 (95%). β symbolizes the probability of committing a type II error, and $1 - \beta$ denotes the probability of making a correct rejection. The power of the test is also known by $1 - \beta$.

No hard and fast rules exist for selecting α and β values. Keeping α, the level of confidence, as small as possible helps reduce the more serious type I error. Unfortunately, as the probability of producing a type I error becomes smaller, the probability of committing a type II error increases, and vice versa. Figure 2 illustrates this concept. As α decreases, the corresponding area under the curve for μ_0 will decrease, and simultaneously the area labeled $1 - \beta$ under the curve for μ will decrease. Hence, the power of the test diminishes.

Fixed Sampling Plan

Increasing n describes the simplest way to reduce α and increase the power of the test. The formula for calculating n is

$$n = \frac{(z_\alpha - z_\beta)^2}{(\mu - \mu_0)^2 / \sigma^2} \tag{6}$$

where z_α identifies the value of z defining the α section for the distribution with μ_0, and z_β corresponds to the value that defines the β section of the distribution with μ. If the hypotheses are nondirectional, z_α is replaced by $z_{\alpha/2}$ for two-tailed values. Consider our standard copper solution labeled 50 ppm. If we choose the minimum acceptable difference between the labeled value and the mean analysis value to be 2 ppm ($\mu - \mu_0$) with an analysis standard deviation of 3 ppm, the number of samples needed with $\alpha/2 = 0.025$ and the power equal to 0.80 computes to

$$n = \frac{[1.96 - (-0.84)]^2}{(2)^2 / (3)^2} = 18$$

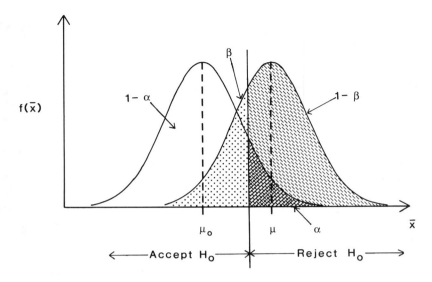

Figure 2 Regions corresponding to type I and type II errors. See text for explanation of symbols.

If one elects to raise the power of the test to 0.95 and keep $\alpha/2 = 0.025$, n increases to 29. In general, for set values of α and $1 - \beta$, the greater the tolerable difference for $\mu - \mu_0$, the less n must be. This procedure for calculating n is known as a *fixed sample plan*.

Problem 1. A standard nickel solution labeled 500 ppm is desired for use in the preparation of a calibration curve for atomic emission. The analysis error is 5.00 standard deviations. If the level of confidence for accepting the labeled concentration is 95% with a selected power of 0.95 for a tolerable difference equal to ±3.00 ppm, how many analyses are needed?

Sometimes the chemist does not know σ or $\mu - \mu_0$. Cohan [3] realized that an *effect size* specified by

$$d = (\mu - \mu_0)/\sigma$$

can be used to represent a relative measure portraying the magnitude of the $\mu - \mu_0$ difference in standard deviation units. In the above example, $d = 2/3 = 0.7$, and the detectable difference desired translates to 0.7 standard deviation. Cohan classifies $d = 0.8$ as a large-effect size, $d = 0.5$ a medium-effect size, and $d = 0.2$ a small-effect size. Hence, one may now specify the minimum detectable $\mu - \mu_0$ difference desired in terms of *effect size*. Correspondingly, Eq. (6) can be expressed as

$$n = (z_\alpha - z_\beta)^2/d^2$$

and calculation of n becomes possible by designating α, β, and d.

Sequential Sampling Plan

If an analysis result is justifiable with a sample size n, the result should be equally valid for a sample size of $n + 1$ or $n - 1$, and so on. Wald [4] developed a sequential sampling plan related to the outcome of an observation. Sequential sampling carries one distinct advantage over fixed sampling: The hypothesis under consideration can be tested after each observation rather than waiting for all n samples to be observed. After each observation, one of three decisions is made: (a) accept the hypothesis; (b) reject the hypothesis; or (c) obtain another observation. Therefore, a decision may be made before all analyses have been performed, perhaps saving time and money. The sequential probability ratio test for normal distributions will be considered here. The three decisions can be represented both mathematically and graphically [2, 4, 5]. Mathematically,

(a) $\sum_{i=1}^{n} (x_i - x_0) \leq \dfrac{\sigma^2}{m_1 - m_0} \ln\left[\dfrac{\beta}{1 - \alpha}\right] + \dfrac{n}{2}(m_1 + m_0)$ or

$\sum_{i=1}^{n} (x_i - x_0) \geq \dfrac{\sigma^2}{m_1 - m_0} \ln\left[\dfrac{\beta}{1 - \alpha}\right] - \dfrac{n}{2}(m_1 + m_0)$

(b) $\sum_{i=1}^{n} (x_i - x_0) \geq \dfrac{\sigma^2}{m_1 - m_0} \ln\left[\dfrac{1 - \beta}{\alpha}\right] + \dfrac{n}{2}(m_1 + m_0)$ or

$\sum_{i=1}^{n} (x_i - x_0) \leq \dfrac{\sigma^2}{m_1 - m_0} \ln\left[\dfrac{1 - \beta}{\alpha}\right] - \dfrac{n}{2}(m_1 + m_0)$

(c) If neither (a) nor (b) is satisfied, continue.

n indicates the current number of samples, x_0 represents the value being tested, and m_1 and m_0 denote the values being tested as H_1 and H_0, respectively.

The first inequalities for (a) and (b) represent one-sided testing, while the additional second inequalities are necessary for two-sided testing. Detection limits, discussed in Section 4.5, portray an example of one-sided testing. Designating a two-sided test indicates that m_1 in the second set of inequalities would simply be the respective value on the low side of m_0. Consider the fixed sampling

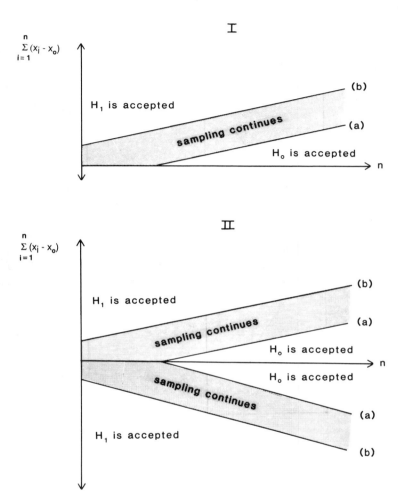

Figure 3 Graphical interpretation of sequential sampling: I, one-sided testing; II, two-sided testing.

illustrated previously where the acceptable difference was set at 2 ppm. For sequential sampling, m_1 and m_0 for the first inequalities in (a) and (b) would be 2 ppm and 0 ppm, respectively, while m_1 and m_0 for the second inequalities would be -2 ppm and again 0 ppm.

Figure 3 shows how (a), (b), and (c) are graphically interpreted for one-sided testing and two-sided testing. [(c) holds when $\sum_{i=1}^{n}(x_i - x_0)$ falls between

CALIBRATION

Table 1 Data for Example 1

x_i	$\sum_{i=1}^{n}[x_i - 500.0]$	n	x_i	$\sum_{i=1}^{n}[x_i - 500.0]$	n
502	2	1	502	43	14
503	7	2	500	43	15
500	7	3	504	47	16
506	13	4	503	50	17
502	15	5	503	53	18
504	19	6	505	58	19
501	20	7	502	60	20
500	20	8	503	63	21
505	25	9	504	67	22
503	28	10	501	68	23
502	30	11	505	73	24
505	35	12	502	75	25
506	41	13			

Using $x_0 = 500$, $m_1 = 3.00$, $m_0 = 0.00$, $\sigma = 5.00$, and $\alpha = \beta = 0.05$, the decision boundaries are (a) $\sum_{i=1}^{n}(x_i - 500) \leq -24.54 + 1.50 n$ or $\sum_{i=1}^{n}(x_i - 500) \geq 24.54 - 1.50 n$
(b) $\sum_{i=1}^{n}(x_1 - 500) \geq 24.54 + 1.50 n$ or $\sum_{i=1}^{n}(x_1 - 500) \leq -24.54 - 1.50 n$

lines (a) and (b).] In either case, the cumulative sum of the deviations from the value specified by x_0 is plotted against the sample number.

Example 1. The chemical analyses listed in Table 1 were performed by atomic emission. When could the analyses stop, and is the solution actually 500 ppm? Compare the number of samples needed here with those for the fixed sample size found in Problem 1.

These boundaries and analyses are plotted in Figure 4, which shows that the testing could terminate after 21 analyses. The null hypothesis stating that the solution is 500.0 ppm must be rejected assuming the operator did not introduce a systematic error. The advantage of sequential sampling becomes evident when compared to the result $n = 36$ calculated by fixed sampling. Figure 4 also shows that a minimum of 14 analyses would be needed before the null hypothesis could be accepted. Since the cumulative deviations from 500.0 ppm are positive, the standard solution is probably greater than the labeled value.

Sampling Quantities and Design

Earlier, Eq. (1) described the overall variance for an analysis result as being the sum of the sampling variance and the analysis variance. Unlike homogeneous populations, heterogeneous populations may develop sampling variances sub-

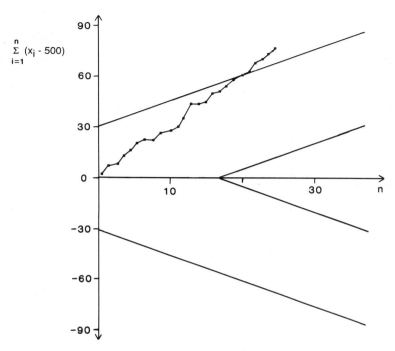

Figure 4 Sequential sampling plan for Example 1.

stantially larger than analysis variances. Since analysis variance usually depends on the analytical method, the sampling plan must use the sample quantity required such that s_s^2 is within a suitable value. For homogeneous populations, or those that can easily be homogenized such as liquids and gases, the sample amount should allow for replicate analyses. Hungerford and Christian [6] applied sampling theory to a simplified model of a homogeneous solution and demonstrated that statistical sampling errors can act as a fundamental limit in detection for dilute solutions.

Determining sample quantities for heterogeneous populations depends on the degree of inhomogeneity with respect to the chemical species of interest, the particle-size variation, and the precision deemed suitable. Liteanu and Rica [2] introduced a technique to evaluate the level of heterogeneity. References 5 and 7–10 review several equations constructed to calculate germane sizes of samples.

The importance of composite sampling resulting from the combination of samples has been clarified with illustrative examples in references 2 and 10–12. Composite sampling may require less analysis time, but information with regard to any variation between individual samples is lost.

Additional Comments on Sampling

Section 4.1 has centered on the proper number of samples or analyses for concentration estimates of the population mean. A further aspect to sampling theory concerns the proper selection of calibration samples for preparation of calibration curves. For a single-component analysis, the calibration sample concentrations should closely bracket the unknown sample concentration. Agterdenbos et al. [13] developed a quantity, the eccentricity (E), to describe how the concentration levels of calibration samples affect the precision of the sample concentration estimate,

$$E = \frac{[\hat{\bar{x}}_s - \bar{x}]^2}{\sum_{i=1}^{n}[x_i - \bar{x}]^2}$$

where $\hat{\bar{x}}_s$ expresses the mean analyte concentration estimate, $\bar{x}$ denotes the mean concentration of the n calibration samples, and x_i symbolizes the concentration of the ith calibration sample. The eccentricity shows that precision is best at $\hat{\bar{x}}_s = \bar{x}$. Similarly, for multicomponent analysis, a corresponding hypervolume exists in the calibration concentration space closely bracketing the unknown sample. Therefore, to obtain precise and accurate concentration predictions, an optimal calibration sample design is essential. Experimental design applicable to this area is discussed in Chapter 6.

4.2 LEAST SQUARES AND LINEAR CALIBRATION

A univariate calibration plot of the instrumental signal, y, versus the concentration of analyte, x, is primarily based on a linear relationship. Responses can be measured with various types of chemical transducers, or sensors, ranging from ion-selective electrodes to particular wavelengths of absorption or emission. The relationship is routinely expressed as

$$y = \beta_0 + \beta_1 x \qquad (7)$$

where β_1 denotes the slope of the calibration curve and β_0 portrays the intercept on the y axis, commonly known as background. The usual procedure consists of four steps: Generate a series of calibration samples containing the analyte at known concentrations, measure their respective instrumental responses, create the calibration plot, and, finally, obtain and transform the unknown sample response to predict concentration. In practice, the data points (x_i, y_i) do not all reside on a single line but rather scatter about the best-fitting line. The scatter is chiefly due to errors in the instrumentally measured y values, and the best fit may be acquired by using the least squares method. Robust methods exist that are less sensitive to outlier distributional assumptions [14].

Univariate Regression and Least Squares

The following discussion assumes that no chemical or physical interferences are present. In addition, utilization of least squares stipulates that significant errors exist only in the dependent variable, y. Further, the variances of y are homoscedastic, that is, the variances are equal, $\sigma_y^2 = N(0, \sigma^2)$. The true model for the ith observation becomes

$$y_i = \beta_0 + \beta_1 x_i + \epsilon_i \tag{8}$$

where ϵ_i denotes the true error defined as the difference between the observed response y_i and the true value, $y_i - (\beta_0 + \beta_1 x_i)$. Unfortunately, we seldom know the true response, and our actual working model is

$$y_i = b_0 + b_1 x_i + e_i \tag{9}$$

with b_0 and b_1 symbolizing estimates of β_0 and β_1. The difference between the observed y_i and the estimated $\hat{y}_i$ [$y_i - \hat{y}_i = y_i - (b_0 + b_1 x_i)$], is termed the observed residual, e_i. Least squares estimates of β_0 and β_1 minimize the sum of the squares of the residuals, $\Sigma_i e_i^2$. If m responses are obtained for m calibration samples, Eq. (9) can be rewritten as

$$\mathbf{y} = \mathbf{Xb} + \mathbf{e} \tag{10}$$

with $\mathbf{y}$ and $\mathbf{e}$ each describing $m \times 1$ vectors, $\mathbf{b}$ signifying the 2×1 calibration vector, and $\mathbf{X}$ designating the $m \times 2$ calibration sample matrix with $m \geq 2$. Using matrix notation, the residuals are estimated from $\mathbf{e} = \mathbf{y} - \hat{\mathbf{y}} = \mathbf{y} - \mathbf{Xb}$, and the inner product matrix $\hat{\mathbf{e}}'\mathbf{e}$ represents the minimized sum of the squared residuals. The prime indicates the transpose matrix operation. The least squares method estimates the β values by

$$\mathbf{b} = [\mathbf{X}'\mathbf{X}]^{-1}\mathbf{X}'\mathbf{y} \tag{11}$$

provided the inverse $[\mathbf{X}'\mathbf{X}]^{-1}$ exists. An example should clarify these matrix equations.

Example 2. A series of copper calibration sample solutions were measured by flame atomic absorption spectrometry, and the following results were obtained.

Concentration (ppm)	3	6	9	12
Absorbance	0.178	0.230	0.311	0.473

CALIBRATION

The calibration parameters can be estimated by

$$\underbrace{\begin{bmatrix} 0.178 \\ 0.230 \\ 0.311 \\ 0.473 \end{bmatrix}}_{\mathbf{y}} = \underbrace{\begin{bmatrix} 1 & 3 \\ 1 & 6 \\ 1 & 9 \\ 1 & 12 \end{bmatrix}}_{\mathbf{X}} \times \underbrace{\begin{bmatrix} b_0 \\ b_1 \end{bmatrix}}_{\mathbf{b}} + \underbrace{\begin{bmatrix} e_1 \\ e_2 \\ e_3 \\ e_4 \end{bmatrix}}_{\mathbf{e}}$$

$\mathbf{b} = [\mathbf{X'X}]^{-1} \mathbf{X'y}$

$$\mathbf{X'X} = \begin{bmatrix} 1 & 1 & 1 & 1 \\ 3 & 6 & 9 & 12 \end{bmatrix} \begin{bmatrix} 1 & 3 \\ 1 & 6 \\ 1 & 9 \\ 1 & 12 \end{bmatrix} = \begin{bmatrix} 4 & 30 \\ 30 & 270 \end{bmatrix}$$

$$[\mathbf{X'X}]^{-1} = \begin{bmatrix} 1.50 & -0.167 \\ -0.167 & 0.022 \end{bmatrix}$$

$$\mathbf{b} = \begin{bmatrix} 1.50 & -0.167 \\ -0.167 & 0.022 \end{bmatrix} \begin{bmatrix} 1 & 1 & 1 & 1 \\ 3 & 6 & 9 & 12 \end{bmatrix} \begin{bmatrix} 0.178 \\ 0.230 \\ 0.317 \\ 0.473 \end{bmatrix}$$

$$= \begin{bmatrix} 0.0540 \\ 0.0314 \end{bmatrix}$$

The calibration model becomes $y_i = 0.0540 + 0.0314 x_i + e_i$ and is plotted in Figure 5. After estimating the parameters and provided the model is adequate, prediction of concentrations for unknown samples becomes possible. If a sample's absorbance was 0.279, the corresponding concentration, $\hat{x}_s$, would be computed as

$$\hat{x}_s = (y_s - b_0)/b_1 = 7.16 \text{ ppm}$$

Curvilinear Regression and Regression Through the Origin

When the simple univariate linear model, Eq. (8), appears to be inadequate, higher order univariate models may be pursued. For example,

$$y_i = \beta_0 + \beta_1 x_i + \beta_2 x_i^2 + \epsilon_i$$

describes a second-order model with one independent variable. Second-order curvilinear regression can be handled as before, with modifications to the di-

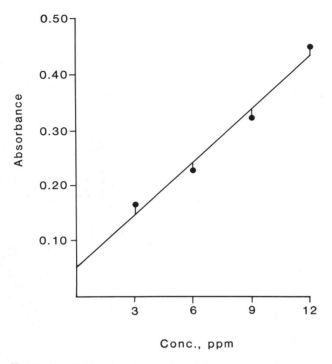

Figure 5 Calibration data and model for Example 2.

mensions of **b** and **X**. Notably, **b** acquires three rows, and **X** has the dimensions of m rows and $2 + 1$ columns, where the $+1$ identifies the intercept. The parameter β_1 is frequently designated the *linear effect parameter*, while β_2 is termed the *curvature effect parameter*. Third-order models are possible as well as higher orders. In general,

$$y_i = \beta_0 + \beta_1 x_i + \beta_2 x_i^2 + \ldots + \beta_p x_i^p$$

where p represents the order of the model. Again, regression can be accomplished as before, with modifications to the dimensions of **b** and **X**. Specifically, **b** acquires $p + 1$ rows, and **X** becomes $m \times (p + 1)$.

Usually, powers higher than the third are not used because interpretation of model parameters becomes difficult. A model of high enough degree can always be established that will fit the data exactly. For example, a univariate data set containing m measurements can be fitted by a model of order $m-1$ that will pass through all m of the measured y values. Hence, the chemist should be suspicious of high-order curvilinear models used to obtain a good fit. Additionally, extrapolation of the curvilinear model beyond the largest value of x_i used to fit the model may not be in agreement with the true situation. For instance, a second-order model, as above, describes a parabola. The parabola eventually starts bending down toward the x axis shortly following the highest calibration sample concentration. With regard to most chemical analyses, this type of curvature is not warranted. Therefore, curvilinear regressions may furnish appropriate fits, but they may deviate in unreasonable directions beyond the range of calibration samples.

Frequently, in calibration, the y value should equal zero when x is zero. With this restriction, the calibration line is constrained to pass through the origin and yields a model with β_0 removed. Removal of b_0 and the column of 1's in $\mathbf{X}$ marks the required alterations to the matrix equations. Caulcutt and Boddy [15] have thoroughly formulated regression through the origin.

Multivariate Regression

Univariate regression is specific for situations where the response depends on one variable or chemical component. With multivariate regression, parameters can be estimated for models where the responses depend on more than one variable or chemical component. Comparable to Eq. (7), the multivariate model for the ith chemical transducer, or sensor, develops into

$$y_i = \beta_0 + \beta_1 x_1 + \beta_2 x_2 + \ldots + \beta_n x_n$$

for n components. Matrix equations analogous to Eqs. (10) and (11) can be used to estimate $\mathbf{b}$, which contains the linear response constants for a particular sensor—specifically, how sensitive that sensor is to each component undergoing calibration. If m calibration samples are used, the dimensions assimilate to $\mathbf{y}$ and $\mathbf{e}$, $m \times 1$; $\mathbf{b}$, $(n + 1) \times 1$; $\mathbf{X}$, $m \times (n + 1)$. Again, the $+1$ indicates the inclusion of an intercept term. For $(\mathbf{X}'\mathbf{X})^{-1}$ to exist, the number of calibration samples must equal or be greater than the number of components plus 1, $m \geq n + 1$. Each row of $\mathbf{X}$ corresponds to the concentrations for each component in a given calibration sample. Complete calibration for concentration predictions of an n-component unknown sample requires several sensors, not just one. Letting p express the number of sensors, $p \geq n$ specifies the required number to solve the following equations.

To formulate these proper expressions for multicomponent analysis using multivariate least squares, we will change from standard statistical notation to a format used more routinely in chemical analysis termed K-matrix. We will use r instead of y, replace b with k, and substitute x with c. In addition, our model will not have an intercept term. As stated earlier, the first step in chemical analysis requires calibration of the chosen p sensors. To do this, we first measure responses at the p sensors for all of the m calibration samples and then form $\mathbf{R}$, an $m \times p$ matrix. Thus, each row of $\mathbf{R}$ contains the responses measured from the p sensors for that calibration sample. Next, the $m \times n$ calibration sample concentration matrix $\mathbf{C}$ is constructed. Each row of $\mathbf{C}$ corresponds to a calibration sample and is composed of the concentration values for the n components present in that calibration sample. Estimation of the $n \times p$, calibration matrix $\mathbf{K}$ involves using

$$\mathbf{R} = \mathbf{CK} + \mathbf{E}$$
$$\hat{\mathbf{K}} = (\mathbf{C'C})^{-1} \mathbf{C'R}$$

where $\mathbf{E}$ specifies the $m \times p$ error matrix. Again, the hat on $\mathbf{K}$ distinguishes an estimated $\mathbf{K}$ as opposed to an exact $\mathbf{K}$. After estimating the calibration matrix, we can proceed with predicting the unknown sample concentration. The $n \times 1$ vector $\hat{\mathbf{c}}_s$ containing the concentration predictions is obtained by measuring the response vector $\mathbf{r}_s$ for the sample and solving

$$\mathbf{r}_s = \hat{\mathbf{K}}' \mathbf{c}_s$$
$$\hat{\mathbf{c}}_s = (\hat{\mathbf{K}} \hat{\mathbf{K}}')^{-1} \hat{\mathbf{K}} \mathbf{r}_s$$

The K-matrix approach applies to any method of chemical analysis that follows Beer's law.

For spectral forms of analyses (IR, UV-Vis, etc.), the K-matrix method can be used as a full-spectrum technique. That is, instrumental responses measured at all wavelengths can be incorporated into the model. For this situation, plots of the rows of $\hat{\mathbf{K}}$ would estimate pure-component spectra at unit concentration. Alternatively, select wavelengths or wavelength ranges can be used with the K-matrix approach. Specifically, wavelengths or spectral regions may be chosen to ensure that only linear responses are measured and used for calibration and analysis. Locating individual wavelengths or ranges requires a criterion to base decisions on and an appropriate method of searching. The reader is directed to references 16-22 for further inquiry.

Unfortunately, a prevalent disadvantage of the K-matrix approach emerges from the condition that all interfering components must be identified and included in the calibration samples at known concentrations. For spectroscopic chemical analysis procedures, select wavelengths or regions may be utilized that single out areas that are not spectrally overlapped by interfering components. The need to include interfering components in calibration samples would be nullified. If select wavelengths are not available and spectral interferences are extensive, errors for predicted concentrations may be severe.

Example 3. Consider the chemical analysis of a sample containing the four components 4-hydroxyacetophenone, 4-hydroxybutyrophenone, 3-acetylaminophenanthrene, and 3-carbethoxy-8-methyl-4-pyrido[1,2-a]pyrimidinone with

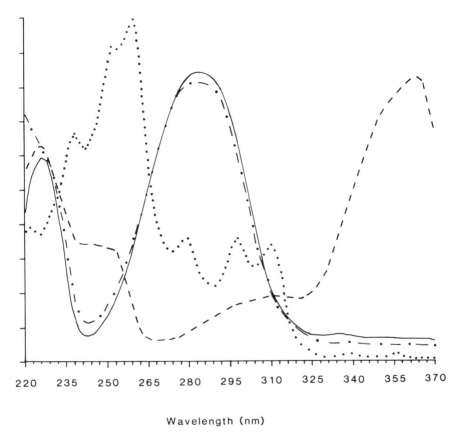

Figure 6 Ultraviolet absorption spectra: (—•—) 3-carbethoxy-8-methyl-4-pyrido[1,2-a]pyrimidinone, (...) 3-acetylaminophenone, (– –) 4-hydroxyacetophenone, and (———) 4-hydroxybutyrophenone.

respective UV absorption spectra illustrated in Figure 6. The use of four calibration samples and every other wavelength (75 wavelengths) generated a 4 × 75 **K** matrix. No attempt was made to select wavelengths that might improve the analysis. As expected, concentration estimates for 4-hydroxyacetophenone and 4-hydroxybutyrophenone listed in Table 2 are severely degraded because of the excessive degree of spectral overlap. Conversely, the concentration estimates for 3-acetylaminophenanthrene and 3-carbethoxy-8-methyl-4-pyrido-[1,2a]pyrimidinone are acceptable because of the prevalent spectral regions with very little spectral overlap for these two components.

Table 2 Comparison of K Matrix with PLS for Components Shown in Figure 6

Component	True (10^{-5})	K Matrix (10^{-5})	PLS (10^{-5})
3-Carbethoxy-8-methyl-4-pyrido[1,2-a]pyrimidinone	1.07	1.06	1.07
3-Acetylaminophenanthrene	7.96	8.06	7.99
4-Hydroxyacetophenone	3.55	4.15	3.92
4-Hydroxybutyrophenone	5.00	4.35	5.46

Alternative Multicomponent Analysis Routines

A few alternative methods of calibration and prediction need mentioning: P matrix, principal component regression (PCR), and partial least squares (PLS). The P-matrix method denotes inverse least squares applied to chemical analysis [23]. The model is given by

$$C = RP + E$$

where C and R are as before, P identifies the $p \times n$ calibration matrix comparable to K, and E portrays an $m \times n$ matrix containing errors associated with the model. The P matrix is estimated by

$$\hat{P} = (R'R)^{-1} R'C$$

and concentration is predicted by

$$\hat{c}'_s = r'_s \hat{P}$$

avoiding the second matrix inversion present in the K-matrix procedure. Equivalent to the K-matrix approach, all interferences must be present in the calibration samples at varying concentration levels. Unlike the K-matrix approach, the concentrations of the interferences need not be known. The major disadvantage stems from the inversion of R, which is mandatory for estimating $\hat{P}$. Particularly, executing this inversion requires that the number of calibration samples be greater than or equal to the number of sensors, $m \geq p$. For example, if 20 sensors are used, at least 20 calibration samples are necessary regardless of the number of analyte concentrations to be predicted. Therefore, proper sensor selection becomes critical for the P-matrix scheme.

Principal component regression (PCR) resembles both the K-matrix and P-matrix methods. Most notably, PCR has full-spectrum capacity and uses an inverse model. Parallel to $C = RP$, the PCR calibration model is

$$C = TQ + E \tag{12}$$

CALIBRATION

where **T** symbolizes the $m \times h$ matrix of scores for the original responses of the calibration samples in a new coordinate system, **Q** identifies the $h \times n$ matrix of regression coefficients, and **E** signifies the matrix of errors associated with the model. Computing the score matrix **T** requires four steps: estimating the p eigenvectors of $\mathbf{R'R}$, retaining h eigenvectors or factors, forming **B** (a $p \times h$ matrix containing the h eigenvectors as columns specifying the new coordinate system or basis set), and computing $\mathbf{T} = \mathbf{RB}$. The regression coefficient matrix **Q** is estimated from a multivariate regression fit of Eq. (12). Analogs to $\hat{\mathbf{c}}_s' = \mathbf{r}_s'\mathbf{P}$, concentration predictions are obtained by computing

$$\hat{\mathbf{c}}_s' = \mathbf{t}_s'\mathbf{Q}$$

where $\mathbf{t}_s'$ designates the score vector in the new coordinate system for the unknown sample responses. The unknown sample score vector is computed from $\mathbf{t}_s' = \mathbf{r}_s'\mathbf{B}$. PCR embraces the advantages of the K-matrix and P-matrix methods without the disadvantages. In particular, unlike the K-matrix approach, PCR achieves accurate and precise concentration predictions in the presence of severe spectral interferences. In addition, since fewer eigenvectors can be used in the calibration, $h \leq p$, noise reduction becomes possible. References 24 and 25 thoroughly communicate the theory of PCR. Recently, Brown and coworkers proposed preprocessing spectra with a Fourier transform [26–28]. Improved results were attained when the raw calibration data were Fourier processed prior to PCR compared to raw data analysis by K matrix, P matrix, PCR, or PLS (described following).

The partial least squares (PLS) approach to calibration and prediction is similar to that of PCR. In PCR, the eigenvectors are computed from calibration spectra independent of concentrations. Therefore, the h retained factors for PCR may not be ideal for concentration predictions. Partial least squares, on the other hand, uses both the calibration responses and concentration information to iteratively determine the factors. The factors are then used to predicate sample concentrations. Since more information is used to build the calibration model, the concentration predictions benefit. For both PCR and PLS, determination of the proper number of significant factors to retain is essential. The theory and methodology of PLS are beyond the scope of this chapter, and the reader may pursue references 29 and 30 for theoretical discussions, reference 31 for a geometric visualization, and references 31–34 for dialogues on the interrelationship between all the calibration and prediction methods mentioned. As expressed in the introduction to this chapter, you need not understand the intrinsic steps of these calibration methods. They are tools that can be easily implemented and are adequately available in commercial software. Table 2 compares results for PLS analysis and K-matrix analysis of Example 3. As expected, PLS significantly improved concentration predictions when severe spectral interferences were present.

Two other multicomponent methods of chemical analysis should be mentioned. These are rank annihilation factor analysis (RAFA) [35], which has been modified to the generalized rank annihilation method [GRAM] [36], and Kalman filtering [37]. Because the emphasis of this chapter is on least squares, the interested reader should consult the indicated references for further information.

Regression Diagnostics

After the model parameters (**b**, **K**, **P**, etc.) have been estimated, the model should be checked for aptness to ensure adequate predictions. Model assumptions, such as linearity, homoscedasticity, or normality of error terms, may not be applicable to the specific data. In addition, unexpected errors can evolve, stemming from outliers, influential data points, or collinearity.

Outliers, usually recognized as anomalous data points with regard to the rest of the data, may or may not invoke an effect on calibration or prediction if removed. Those outliers whose removal dramatically affect results are said to be *influential*.

Collinearity defines the degree of similarity in calibration responses. In terms of spectroscopy, collinearity pertains to the degree of spectral overlap and similarity between calibration spectra. When collinearity becomes a problem, the K-matrix and P-matrix procedures suffer. This occurs because they require matrix inversions of $\hat{K}\hat{K}'$ and $R'R$, respectively. As a consequence, the inversions are numerically unstable when serious collinearity exists. Matrices of this form are termed *ill-conditioned* [38, 39]. A basic introduction to some of the more simple diagnostic procedures is presented here.

Model Departures and Outliers

Assessment of model departures and the presence of potential outliers can be interpreted from residual plots. For example, if all assumptions about the model proposed in Eq. (8) are correct, a plot of residuals against the independent variable should show a horizontal band. Figure 7a illustrates this. A plot similar to Figure 7b implies violation of homoscedasticity. Specifically, the error variances are not all equal but instead increase with each increment of an independent value. Heteroscedasticity describes this type of phenomenon, and weighted least squares provides a feasible remedy [15, 40]. Variance transformation of the data offers another viable route [40]. Figure 7c characterizes model departure and suggests data transformation or the use of extra terms such as squares or cross products. Section 4.4 develops data transformations; curvilinear calibration pertaining to the inclusion of extra terms was discussed earlier in this section. The residuals can also be plotted versus the dependent variable.

CALIBRATION

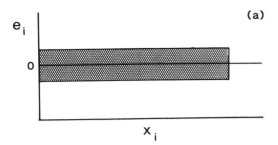

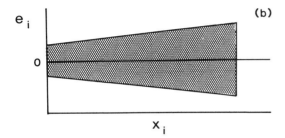

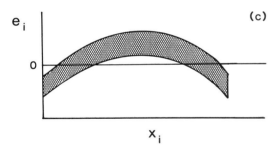

Figure 7 Possible residual plots.

Example 4. The model $y = \beta_0 + \beta_1 x$ was used to fit the data plotted in Figure 8. The fitted calibration curve is also plotted. Figure 9 shows the residuals plotted versus the independent variable. This residual plot resembles the inverse of Figure 7c, suggesting that the linear model is not appropriate and additional terms may be added or the data can be transformed.

Recalling that $\mathbf{y} - \hat{\mathbf{y}}$ estimates residuals, then

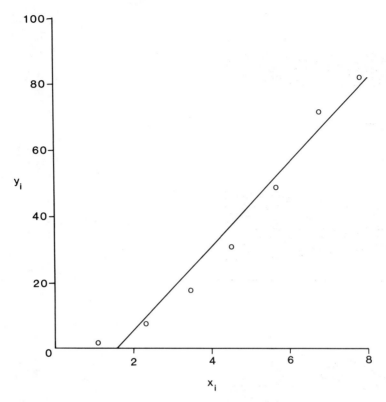

Figure 8 Least squares fit to (1.10, 1.59), (2.32, 7.61), (3.46, 17.6), (4.52, 31.0), (5.62, 48.6), (6.74, 71.5), and (7.81, 81.7). Estimated model: $y = -20.50 + 12.8x$. See Example 4.

$$e = y - Xb$$
$$= y - X[X'X]^{-1} X'y$$
$$= [I - H]y$$

where $H = X[X'X]^{-1}X'$ and is called the *hat matrix* [40, 41]. Hoaglin and Welsch [41] have designated the diagonal elements of H, h_{ii}, as leverage for the ith observation (calibration sample in our case) on the corresponding fitted value. Calibration samples with a large h_{ii} ($0 \leq h_{ii} \leq 1$) are potentially the most influential. More complete indicators of influence requires consideration of leverage values from X and the information in y. References 41 and 42 contain the necessary information to get started. Many of these measures have been modified for PCR [24, 43, 44] and PLS [45].

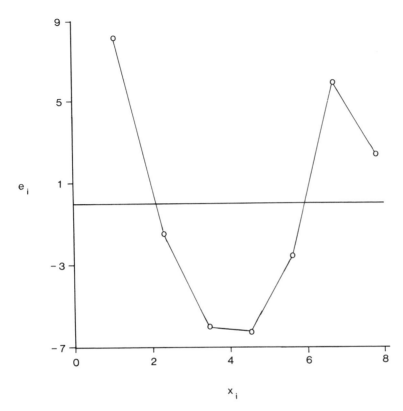

Figure 9 Residuals for data and model from Figure 8.

Collinearity

As stated earlier, severe collinearity can seriously degrade concentration predictions for the K- and P-matrix methods. Before attempting a regression, one may assess the potential extent of degradation due to collinearity. References 39, 42, and 46 review most of the methods for diagnosing collinearity and the extent of involvement by each component. Section 4.6 discusses some of these methods. Generally, collinearity is not a problem with biased regression techniques such as PCR and PLS. Other biased regression estimators exist, and reference 47 includes additional information about them.

In summary, identifying outliers and influential observations is just as important as obtaining the regression results. Moreover, after influential observations have been pinpointed, a decision as what to do with them must be made. Robust estimations are available that automatically minimize influential observations by downweighting them [14]. A critical examination of observations significantly downweighted is still essential.

4.3 CONFIDENCE INTERVALS

After calculating calibration parameters for a univariate model, it is worthwhile to examine the errors existing in b and establish confidence intervals. Equations introduced here apply to simple and higher order models. All t values should be acquired for $n-p$ degrees of freedom. For the simple model, Eq. (8), this gives $n-2$ degrees of freedom. If one uses weighted least squares or regression through the origin, the following developed confidence intervals are not applicable. Caulcutt and Boddy [15] furnish the necessary information for constructing confidence intervals for both situations, while Massart et al. [48] provide a matrix algebra approach for weighted least squares.

Confidence Intervals for the Intercept and Slope

The variance–covariance matrix for **b** is represented by $\sigma^2(\mathbf{b})$ and expressed as

$$\sigma^2(\mathbf{b}) = \begin{bmatrix} \sigma^2(b_0) & \sigma(b_0, b_1) \\ \sigma(b_1, b_0) & \sigma^2(b_1) \end{bmatrix} = \sigma_y^2 [\mathbf{X}'\mathbf{X}]^{-1}$$

The diagonal terms describe the variances for **b**, while the off-diagonal terms convey the covariances between the parameters contriving **b**, and σ_y^2 expresses the homoscedastic calibration model variance. The variance–covariance matrix can be estimated by

$$\mathbf{s}^2(\mathbf{b}) = s_y^2 [\mathbf{X}'\mathbf{X}]^{-1} \tag{13}$$

where s_y^2 represents the estimated calibration model variance. The calibration model variance can be estimated by one of two approaches: replication of response measurements and using Eq. (4) or using the sum of the squares of the errors (residuals), abbreviated SSE. The latter is introduced here. Recalling our definition of residuals,

$$\text{SSE} = \mathbf{e}'\mathbf{e} = [\mathbf{y} - \hat{\mathbf{y}}]'[\mathbf{y} - \hat{\mathbf{y}}] = \mathbf{y}'\mathbf{y} - \mathbf{b}'\mathbf{X}'\mathbf{y}$$

s_y^2 can be estimated using the equation

$$s_y^2 = \text{SSE}/(n - 2)$$

for $n-2$ degrees of freedom. The respective confidence intervals are written as

$$\beta_0 = b_0 \pm t s(b_0) \tag{14}$$

and

CALIBRATION

$$\beta_1 = b_1 \pm ts(b_1) \tag{15}$$

obtaining t from a t table for $n-2$ degrees of freedom.

Example 5. Returning to Example 2, SSE = 0.00316, producing $s_y^2 = 1.90 \times 10^{-4}$. Using this information and Eq. (13), the variance–covariance matrix becomes

$$s^2(\mathbf{b}) = \begin{bmatrix} 2.85 \times 10^{-4} & -3.18 \times 10^{-5} \\ -3.18 \times 10^{-5} & 4.23 \times 10^{-6} \end{bmatrix}$$

generating the confidence intervals

$$\beta_0 = 0.0540 \pm t(1.69 \times 10^{-2})$$
$$\beta_1 = 0.0314 \pm t(2.06 \times 10^{-3})$$

At the 95% confidence level, the confidence interval for b_0 computes to 0.0540 ± 0.0727, suggesting a possible zero intercept. Consequently, a model with the calibration line passing through the origin becomes plausible.

Confidence Intervals for the Calibration Line

The predicted value $\hat{y} = \mathbf{x}_0'\mathbf{b}$ for a given concentration level x_0 estimates $y = \mathbf{x}_0'\boldsymbol{\beta}$. In other words, the response predicted for a given concentration depends on the least squares estimate of $\mathbf{b}$. The confidence interval for the *true mean value* of the distribution for y at x_0 with the variance for $\hat{y}$ determined by $s_y^2[\mathbf{x}_0'(\mathbf{X}'\mathbf{X})^{-1}\mathbf{x}_0]$ is computed according to

$$\hat{y} \pm ts_y\sqrt{\mathbf{x}_0'(\mathbf{X}'\mathbf{X})^{-1}\mathbf{x}_0} \tag{16}$$

Example 6. Using model parameters acquired in Example 2, the confidence interval for the *true mean* at $x_0 = 7$ ppm computes to

$$\begin{bmatrix} 1 & 7 \end{bmatrix} \begin{bmatrix} 0.0540 \\ 0.0314 \end{bmatrix} \pm t(1.38 \times 10^{-2}) \sqrt{\begin{bmatrix} 1 & 7 \end{bmatrix} \begin{bmatrix} 1.50 & -0.167 \\ -0.167 & 0.022 \end{bmatrix} \begin{bmatrix} 1 \\ 7 \end{bmatrix}}$$

For

$$0.274 \pm t0.0068$$

Successive calculations of Eq. (16) for various values of x_0 at the 90% confidence level generated the inner confidence band drawn in Figure 10.

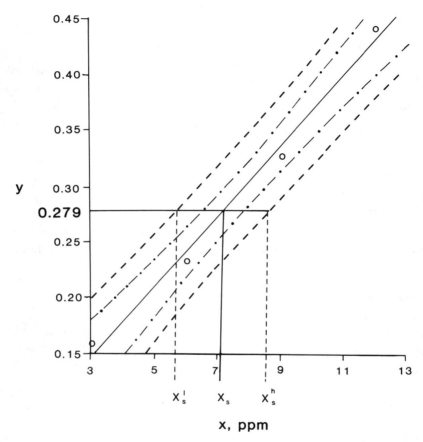

Figure 10 Least squares fit and confidence intervals for data from Example 2. See text for discussion of confidence intervals.

The confidence interval for the predicted mean of m responses measured at a set x_0 is given by

$$\hat{y} \pm t s_y \sqrt{\frac{1}{m} + \mathbf{x}_0'(\mathbf{X}'\mathbf{X})^{-1}\mathbf{x}_0} \tag{17}$$

If one wishes to predict only one response, $m = 1$, the widest confidence interval is obtained. The outer confidence band plotted in Figure 10 depicts $m = 1$ for the 90% confidence level. The plotted confidence intervals clearly show how the upper and lower limits change as the position of x_0 changes.

CALIBRATION

Effect of X

Confidence intervals for regression parameters [Eqs. (13)–(15)] and calibration lines [Eqs. (16) and (17)] are all dependent on $(X'X)^{-1}$. Among other things, the analyst needs to consider the number, spacing, extreme concentrations, and level of replication for the calibration samples. Depending on the purpose, different answers exist.

In general, one acquires the most precise estimate of β_0 by collecting multiple measurements at the lowest allowable level and one measurement at the highest allowable level. The most precise estimate of β_1 takes place when half the calibration solutions have the lowest possible concentration levels and the other half have the highest possible levels. Superior precision for estimating a predicted response becomes attainable when concentration levels are such that $\bar{x} = x_0$. At the same time, precision is reduced when concentration levels are distributed far from $\bar{x}$, similar to β_1. In practice, calibration samples are usually evenly distributed, ensuring sufficient bracketing of the unknown sample concentration. Chapter 6 describes additional information on experimental design, a procedure for achieving these goals.

Confidence Interval for Predicted Concentration

Figure 10 displays the effect of the outer regression band on the concentration prediction for an unknown sample, x_s. As x_s advances closer to the mean value of the calibration samples, $\bar{x}$, the confidence interval decreases. The outer confidence interval for x_s producing response y_s is approximated by

$$x_s \pm t \frac{s_y}{b_1} \left[\frac{1}{m} + \frac{1}{n} + \frac{(y_s - \bar{y})^2}{b_1^2 \sum_i (x_i - \bar{x})^2} \right]^{1/2} \tag{18}$$

where m indicates the number of replicate measurements acquired for the unknown sample, n expresses the number of calibration samples used to derive the calibration curve, and $\bar{y}$ represents the average of the calibration sample responses. Equation (18) reveals numerous approaches for minimizing confidence intervals. Increasing the number of calibration samples decreases the confidence interval, because n will be larger. Simultaneously, the number of degrees of freedom $(n-2)$ increases, thereby reducing t. As the mean calibration sample response approaches y_s, the interval becomes smaller. The interval becomes most narrow when calibration sample concentrations are as distant as possible from $\bar{x}$, maximizing $\Sigma(x_i - \bar{x})^2$. Increasing the number of repeat determinations for an unknown sample and using the average response for these samples also decreases the interval. If the sample response 0.279 used earlier were actually the average of three separate sample preparations and measurements, the confidence interval would be $7.16 \pm t0.336$ compared to $7.16 \pm t0.491$ for only one measurement.

Multivariate Confidence Intervals

Estimating errors and confidence intervals for predicted concentrations acquired through a multivariate analysis is difficult because of the nonlinear propagation of errors [49]. For the K-matrix approach, a rough approximation of the confidence intervals is possible if one knows **K** accurately. Parallel to Eq. (13), the variance–covariance matrix for $\hat{c}_s$ computes as

$$s^2(\hat{c}_s) = s_{r_s}^2 [\hat{K}\hat{K}']^{-1}$$

with $s_{r_s}^2$ symbolizing the estimated calibration model variance. Estimation of the experimental error can be calculated from

$$s_{r_s}^2 = SSE/(p - n)$$

for $p-n$ degrees of freedom. The SSE term is for the residuals of $\mathbf{r}_s$, calculated by $\mathbf{r}_s - \hat{\mathbf{r}}_s$. Confidence intervals are written as

$$c_s = \hat{c}_s \pm ts(\hat{c}_s)$$

More accurate confidence intervals can be estimated but require additional computations. Kalivas and Lang [50] approached the problem geometrically, whereas Lorber and Kowalski [49] proposed a statistical solution. Both methods are beyond the scope of this chapter, and the interested reader should examine the respective references.

As in univariate analysis, multivariate analysis confidence intervals are affected by the design of calibration samples. References 51–55 should be consulted for further details on the effects and procedural steps for optimal calibration sample selection.

4.4 NONLINEAR CALIBRATION

Since Chapter 5 offers an in-depth discussion of nonlinear regression analysis, only a few brief items are assembled here. If a nonlinear calibration plot is obtained or indicated by a residual plot, various options exist. One is to restrict analyses to linear ranges. Specifically, linear approximations over restricted ranges may form competent models. Of course, separate ranges demand different linear models, and predictions outside an approximated linear range are not meaningful.

Alternatively, if the nonlinear calibration model is known, transformation to a linear model becomes possible. For example, the model $y = ax^b$ appears

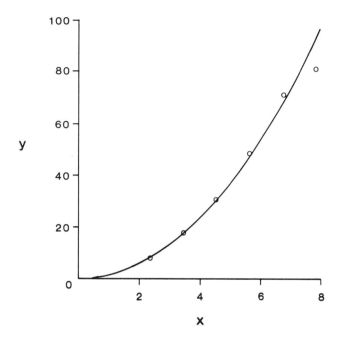

Figure 11 Data from Figure 8 plotted using $y = 1.98x^{2.08}$ ($y = ax^b$).

often in spectroscopy. Implementing a logarithmic transformation produces $\log y = \log a + b \log x$ or $\log y = b_0 + b_1 \log x$. Figure 9 displays residuals from using the linear model $y = b_0 + b_1 x$ to fit the data plotted in Figure 8. The residual plot clearly indicates the necessity for data transformation. Figure 11 contains the same data using $y = ax^b$. A difficulty encountered with fitting this or any other nonlinear model directly is that $\sum_{i=1}^{n} [y_i - ax^b]^2$ must be minimized. Performing this task creates a series of equations not linear in parameters requiring iterative procedures for solutions. Chapter 12 of reference 40 provides an introduction to such methods of nonlinear least squares. However, executing a linear transformation permits the use of linear least squares. Figure 12 shows the data using the log-transformed model.

One should exercise caution when applying linear transformations. Parameter estimates obtained from transformed models are not the same as those acquired from the original model by nonlinear least squares. For example, linear least squares assumes that the original variables are normally distributed. Transformation may invalidate this conjecture. Additionally, transformation may nullify the homoscedasticity of variance assumption.

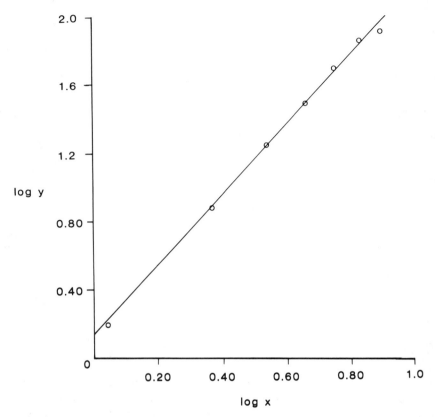

Figure 12 Least squares fit to the log transformed data of Figure 8. Estimated model: log y = 0.131 + 2.05 log x.

Mark [56] has catalogued the effects of nonlinearity of multivariate calibrations. Included in his discussion are cases when curvature exists in the independent or dependent variables, or both.

4.5 SENSITIVITY AND LIMITS OF DETECTION

Sensitivity and detection limits are often confused and thought to be synonymous. For example, a methodology with a low detection limit for a particular component is frequently said to be sensitive towards that component. Sensitivity actually denotes the slope of a calibration curve and describes the change in

CALIBRATION

response per unit change in concentration. A fair amount of confusion also exists for reported detection limits due to an assortment of calculation methods [57–62]. The IUPAC definition is "the limit of detection, expressed as the concentration, c_L, or the quantity, q_L, is derived from the smallest measure, x_L, that can be detected with reasonable certainty for a given analytical procedure" [59]. Apparently, much of the disparity stems from the interpretation of *reasonable certainty*. This section will emphasize the approach of Currie [57], Kaiser [60], Boumans [61], and Hubaux and Vos [62] for detection limits.

Sensitivity

The sensitivity for univariate calibration denotes the slope of the calibration curve, dy/dx. Confidence intervals determined for sample concentration predictions are strongly influenced by sensitivities. Specifically, Eq. (18) reveals that if the sensitivity (b_1) can be increased, the confidence interval decreases.

The definition of sensitivity becomes more complicated for multivariate calibration. In the K-matrix case, numerous sensitivity coefficients exist because each element of $\mathbf{K}$ represents the slope of a calibration curve. Junker and Bergman [63] generalized Kaiser's [64] definition of sensitivity for multicomponent systems to be the square root of the determinant of $\mathbf{KK'}$:

$$\text{Sensitivity} = (|\mathbf{KK'}|)^{1/2}$$

The generalization allowed more sensors than components to be included in the model.

Figure 6 contains UV-Vis spectra of four components. The spectra of 3-carbethoxy-8-methyl-4-pyrido[1,2-*a*]pyrimidinone (component 1) and 3-acetylaminophenanthrene (component 2) differ distinctly. Conversely, the spectra for 4-hydroxyacetophenone (component 3) and 4-hydroxybutyrophenone (component 4) are similar and almost exactly collinear. If one uses every other sensor, the sensitivity for a system composed of components 1 and 2 is 14.7. Table 3 shows the sensitivity first increasing to 39.4 with the addition of component 3 and then decreasing to 10.2 when component 4 is included. Apparently, the sensitivity of a multicomponent system may decrease as the degree of spectral overlap increases.

A more quantitative measure of sensitivity would compute values for each component present in a sample. Lorber recently derived a series of figures of merits, including sensitivity, for multicomponent systems [65]. Sensitivity for the *j*th component is defined to be that part of the component's spectrum orthogonal to the other sample component's spectra ratioed to the *j*th component's concentration. Mathematically,

$$\text{SEN}_j = \|\mathbf{a}_j^*\| / c_j^0$$

Table 3 Comparisons of Figures of Merit for Components Shown in Figure 6

Mixture components[a]	$(|KK'|)^{1/2}$	SEN_j	cond(K)	SEL_j
1		60.4		0.886
2	14.7	103	1.82	0.886
1		59.4		0.870
2	39.4	84.1	2.75	0.727
3		13.4		0.742
1		59.3		0.869
2		67.6		0.584
3	10.2	1.01	44.8	0.555
4		0.861		0.536
3		1.22		0.672
4	1.19	1.08	30.6	0.672

[a] 1 = 3-Carbethoxy-8-methyl-4-pyrido[1,2-a]pyrimidinone; 2 = 3-acetylaminophenanthrene; 3 = 4-hydroxyacetophenone; 4 = 4-hydroxybutyrophenone.

where c_j^0 designates the calibration concentration of the jth component and a_j^* specifies that part of the spectrum for component j orthogonal to the other components. The $\|\ \|$ signifies vector norm, the square root of the sum of the squares of its elements. Table 3 lists sensitivities for various combinations of the components shown in Figure 6. Once more, the sensitivities are observed to decrease as the level of spectral interference increases. In an earlier paper [66], Morgan used a similar approach in discussing sensitivity losses due to spectral interferences.

Univariate Limit of Detection

Often, trace level analyses must be performed. Prior to transforming a component's signal to concentration, one needs to decide if the measured signal is significantly above the background. Unfortunately, disagreement prevails in the literature on how to define "significantly above the background." Additionally, various terms have been specified for the same definitions. The terminology introduced by Currie [57] will be used here.

Decision Limit

The decision limit corresponds to the critical level for a signal, y_c, at which an observed signal can be distinguished reliably from the background. If interferences are absent and measurement errors for the blank and sample containing the component follow normal distributions, the distributions can be represented

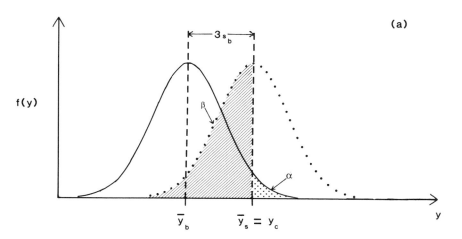

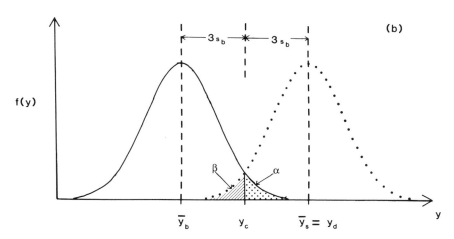

Figure 13 Graphical representation. (a) Decision limits with $\alpha = 0.0013$, $z_\alpha = 3.0$, and $\beta = 0.50$; (b) detection limits with $\alpha = 0.0013$, $z_\alpha = 3.0$, and $\beta = 0.0013$.

as in Figure 13. $\bar{y}_b$ and $\bar{y}_s$ symbolize the blank and sample measurement means, respectively, while s_b and s_s represent the corresponding standard deviations. The distributions drawn in Figure 13 are for $s_b = s_s$, which is usually true at trace levels. If y_s and y_b specify the signals measured for a sample and blank, respectively, then in terms of hypothesis testing, the null hypothesis can be stated as:

The component is not present or The sample signal does not significantly differ from the blank (H_0: $y_s = y_b$); the alternative hypothesis becomes: The component is present or The sample signal is significantly different than the blank (H_1: $y_s > y_b$). Accepting the alternative hypothesis when the sample component is not present invokes a type I error with probability α. Accepting the null hypothesis when the sample is present renders a type II error with probability β.

Acceptance or rejection of the null hypothesis is based on a set critical level for the sample signal (y_c), commonly expressed as

$$y_c = \bar{y}_b + k_c s_b$$

where k_c specifies a number governed by the risk one wishes to accept for a type I error. Determining $\bar{y}_b$ and s_b from many measurements implies suitable estimates of the population values μ_b and σ_b. Therefore, z_α can be used for k_c. If we let the risk of a type I error be $\alpha = 0.0013$, the critical z value corresponds to 3.00 for the $1-\alpha = 99.87\%$ confidence level. The decision limit becomes

$$y_c = \mu_b + z_\alpha \sigma_b = \mu_b + 3\sigma_b$$

With this risk level, a 0.13% chance exists that a sample without a component present would be interpreted as having the component present. Unfortunately, the chance of making a type II error becomes $\beta = 0.50$, expressing a risk of failing to detect the component 50% of the time. Figure 13a graphically shows the problem.

In practice, only a limited number of measurements are made to compute $\bar{y}_b$ and s_b. The constant k_c should then be replaced by the appropriate t value with the proper degrees of freedom.

Detection Limit

The detection limit, y_d, represents the signal level that can be relied upon to imply detection. To avoid the large β value for decision limits, a larger critical signal becomes necessary. The blank and sample signals would have analogous distributions as earlier, but the sample would be centered around a greater value denoted as y_d. Choosing y_d such that $\alpha = \beta = 0.0013$ substantially reduces the probability of obtaining a measurement for a sample below y_c. Figure 13b illustrates this. The signal level at which this occurs identifies the detection limit and is expressed as

$$y_d = y_c + 3\sigma_b = \mu_b + 6\sigma_b$$

Thus, requiring a larger critical signal level considerably diminishes the chance of making a type II error.

An alternative definition for the detection limit, prevalent in the literature, substitutes 3.00 for 6.00. This formulates a decision limit of $y_c = \mu_b + 1.5\sigma_b$. For this alternative definition, the probabilities of making type I and II

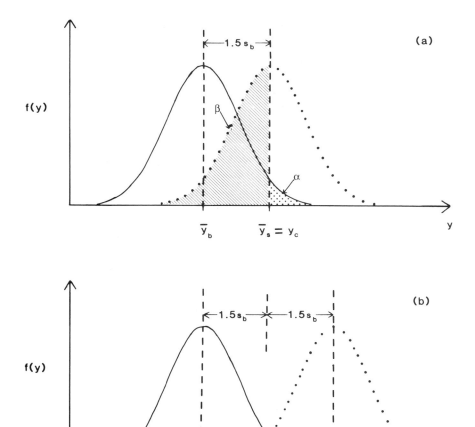

Figure 14 Graphical representation. (a) Decision limits with $\alpha = 0.067$, $z_\alpha = 1.5$, and $\beta = 0.50$; (b) detection limits with $\alpha = 0.067$, $z_\alpha = 1.5$, and $\beta = 0.067$.

detection limit errors have modified to $\alpha = \beta = 0.067$ for a $1-\alpha = 93.30\%$ confidence level. Figure 14 depicts the situation revealing that this definition of detection limit runs a greater risk of making errors. Apparently, numerous definitions for detection limit are possible, with each definition depending on the designated level of confidence. For example, at the 95% confidence level, $\alpha = 0.05$, $z_\alpha = 1.645$, $y_c = \mu_b + 1.645\sigma_b$, and $y_d + 3.29\sigma_b$. Therefore, reported detection limits should be accompanied by the level of significance selected. In general, the greater the confidence level, the larger the detection limit.

Similar to the decision limit, appropriate t values for the proper degrees of freedom should be used if a small number of measurements are made. Hubaux and Vos [62] showed how to estimate the detection limit based on confidence intervals computed for the calibration curve. Their method of estimation is directly comparable to the one presented here.

Example 7. Using the 99.87% confidence level, the detection limit for the calibration data given in Example 2 is calculated by the following. The detection limit for the signal is $y_d = \mu_b + 6\sigma_b$. The data in Example 2 show only four measurements, none of which were for a blank. Thus, the values for b_0 and s_y are used as estimates for μ_b and σ_b, respectively. The value of y_d is computed to be $0.0540 + 6(1.38 \times 10^{-2}) = 0.137$. The concentration detection limit, x_d, is calculated from the calibration model; $y_d = 0.0540 + 0.0314x_d$. This determines $x_d = 2.64$ ppm. Alternatively, x_d could be estimated as

$$x_d = z_\alpha \sigma_b / b_1$$

where b_1 is the slope of the calibration curve. This discloses that by enhancing the sensitivity, the detection limit can be lowered.

Problem 2. Use the 93.30% confidence level to calculate the detection limit for the calibration data given in Example 2. Compare your answer with Example 7.

Problem 3. A sample that yields measurements greater than the blank and less than the detection limit may still contain the component. Is copper present in a sample similar to that of Example 2 if the following measurements were acquired? Use the 93.3% level of confidence.

y_i	n	y_i	n
0.068	1	0.082	7
0.075	2	0.084	8
0.084	3	0.074	9
0.074	4	0.081	10
0.069	5	0.079	11
0.079	6		

Determination Limit

The determination limit y_q designates the signal level at which an acceptable quantitative analysis can be made. A value of

CALIBRATION

$$y_q = \bar{y}_s + 10\sigma_b$$

is typically used for reported determination limits.

Multivariate Detection Limits
Little work has been performed toward a multivariate model for detection limit estimators. Lorber [65] developed a model for multicomponent systems based on net analyte signals. Delaney [67] used a principal component analysis approach. Garner and Robertson [68] suggested using discriminant analysis.

4.6 INTERFERENCE EFFECTS

Interferences are common and severe problems that can render chemical analyses invalid. In general, interferences are classified as physical, chemical, or spectral. *Physical interferences* are caused by the effects of physical properties of the sample solution on the physical processes involved in the analytical measurements. The viscosity, surface tension, and vapor pressure of the sample solution are physical properties that commonly cause interferences in atomic absorption (AA) and atomic emission (AE). *Chemical interferences* are those chemical interactions between the analyte and other substances present in the sample that can influence the analyte signal. A common example in AA and AE is the effect of an easily ionized concomitant element, which can cause signal enhancement by suppressing ionization of the analyte. *Spectral interferences* are those that arise when a sensor is not completely selective for the analyte and are quite common in most spectroscopic methods of analysis.

Physical and chemical effects can be combined and identified as sample matrix effects. Interferences can therefore be considered to have two sources: spectral and sample matrix effects. Spectral interferences cause parallel shifts in calibration curves, whereas sample matrix effects change slopes. This section will concentrate on assessing spectral interferences, and Section 4.7 will emphasize matrix effects and correction procedures.

Selectivity

Example 3 and Table 2 presented results demonstrating the problem of inaccurate concentration predictions when spectral interferences are severe. The degree of spectral interferences, and hence the potential harm, can be estimated before quantization occurs. Selectivity describes the degree of spectral interferences, and several measures have been proposed [39, 46, 65, 69].

The condition number of $\mathbf{KK}'$, cond(K), first introduced to analytical chemistry by Jochum et al. [38], has been used as a selectivity measurement for expressing the degree of spectral overlap. It is computed as the largest singular value of $\mathbf{K}$ divided by the smallest. Spectra with cond(K) = 1 represent complete orthogonality, while cond(K) > 1 indicates spectral overlap at some of the p wavelengths. As cond(K) increases with the extent of spectral overlap, concentration predictions tend to numerically degrade. Table 3 displays cond(K) for different combinations of the spectra illustrated in Figure 6. The largest cond(K) arises when all components are present. Further investigation shows that the spectral overlap of components 3 and 4 influenced this cond(K) the greatest. Unfortunately, the usefulness of cond(K) is limited because it does not disclose which components will have degraded concentration predictions. Table 2 shows that acceptable results are possible for components 1 and 2 in a mixture containing all four. Cond(K) revealed that a problem existed but did not indicate which components were the problem. Therefore, the condition number is only a convenient indicator of the overall quality of a chemical system. It should be noted that a system with suitable selectivity may not have appropriate sensitivity [47, 69–71]. Thus, the two figures of merit, cond(K) and $(|\mathbf{KK'}|)^{1/2}$, should not be treated as synonymous information.

Lorber [65] proposed a new mathematical representation of selectivity for each component. Similar to sensitivity, the selectivity for the jth component is the part of the component's spectrum that is orthogonal to the other spectra ratioed to its spectrum $\mathbf{a}_j$.

$$\text{SEL}_j = \| \mathbf{a}_j^* \| / \| \mathbf{a}_j \|$$

A selectivity value for a given component describes the overall collinearity between that component and the rest. Morgan [66] used a similar approach in expressing losses due to spectral interferences. Table 3 contains computed selectivity values for combinations of the spectra presented in Figure 6.

Kalivas [46] recently presented a new method of evaluating spectra for selectivity. The basis of the method stems from a regression diagnostic tool introduced by Belsley et al. [42] termed *variance decomposition*. Condition indexes are defined as

$$\mu_{\max}/\mu_i \quad i = 1, 2, \ldots, n$$

where $\mu_{\max}$ depicts the maximum singular value of $\mathbf{K}$ and μ_i symbolizes the ith singular value for n components. The largest condition index is cond(K). The condition indexes are tabulated along with the corresponding variance-decomposition matrix $\mathbf{\Pi}$. References 42 and 46 should be consulted for details on calculating $\mathbf{\Pi}$. Listed in Table 4 are the results of variance decomposition of the spectra in Figure 6. Note that each column of a $\mathbf{\Pi}$ matrix should sum to 1. The diagnostic procedure suggests using condition indexes equal to 5 for weak spectral dependencies and equal to 10–100 for moderate to strong relationships. The second $\mathbf{\Pi}$ matrix shows the weak spectral overlap problem for components 2 and 3. The third $\mathbf{\Pi}$ matrix has a large condition index in the fourth row. Inspection of the fourth row of $\mathbf{\Pi}$ exposes components 3 and 4 as the specific components

CALIBRATION

Table 4 Variance Decomposition of Figure 6

\multicolumn{4}{c}{$\pi_{j,i}$ for component[a]}	Condition
1	2	3	4	index
0.377	0.377			1.00
0.623	0.623			1.82
0.168	0.202	0.198		1.00
0.824	0.094	0.179		1.82
0.008	0.704	0.623		2.75
0.074	0.103	0.014	0.013	1.00
0.083	0.868	0.013	0.008	2.01
0.842	0.001	0.009	0.008	2.66
0.000	0.028	0.963	0.970	44.8
		0.033	0.033	1.00
		0.967	0.967	30.6

[a] 1 = 3-Carbethoxy-8-methyl-4-pyrido[1,2-a]pyrimidinone; 2 = 3-acetylaminophenanthrene; 3 = 4-hydroxyacetophenone; 4 = 4-hydroxybutyrophenone.

involved in spectral overlap. Components 1 and 2 would be diagnosed as having suitable selectivity. These methods of assessing selectivity, in addition to others, are explicitly compared in reference 46.

4.7 STANDARD ADDITION

Univariate Standard Addition

The standard addition method (SAM) can be used to correct for sample matrix effects. Procedural steps encompass dividing the sample into several equal-volume aliquots, adding increasing amounts of the proper standard to all aliquots except one, diluting each aliquot to the same volume, measuring instrument signals, and plotting the results as shown in Figure 15. This plot is somewhat different from the one introduced in Section 4.2. However, the same least squares approach can be used by defining new model parameters. From Eq. (8), y_i now represents the response after the ith addition, b_0 converts to the response for the sample solution with no additions, x becomes the concentration of standard added in each addition, and b_1 remains the slope. When $y_i = 0$, the sample concentration prediction shown in Figure 15 computes to $x_s = b_0/b_1$.

Several investigators have advocated performing the standard additions without diluting to a constant volume. Multiplying the measured responses by a

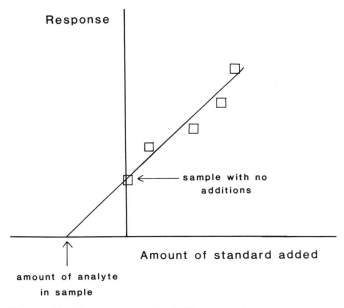

Figure 15 Univariate standard addition method.

ratio of total to initial volume accomplishes a correction for dilution. Unfortunately, the matrix concentration is diluted by the additions, creating a nonlinear matrix effect that may or may not be transformed into a linear effect by volume correction. Kalivas [72] has shown the critical importance of constant volume for the SAM. Maintaining constant volume ensures that any sample matrix effects should be the same throughout, and, in effect, a calibration plot is prepared with *exact* matrix matching of the sample and standards. Other requirements for the SAM are that the response for the analyte should be zero when its concentration equals zero, the response is a linear function of the concentration of the analyte, and sample matrix effects are independent of the ratio of the analyte and matrix. It is important to note that the SAM cannot be used when spectral interferences are present.

Analogous to concentration predictions from a least squares fitted model, concentration predictions for the SAM contain errors, and confidence intervals can be reported. The interval is computed by

$$x_s \pm t \frac{s_y}{b_1} \left[\frac{1}{n} + \frac{\bar{y}^2}{b_1^2 \, \Sigma_i (x_i - \bar{x})^2} \right]^{1/2} \tag{19}$$

CALIBRATION

for n total solutions and $n-2$ degrees of freedom.

Example 8. Aliquots of 15.00 mL containing lake water were delivered into five 50.0-mL volumetric flasks. One flask was diluted to the mark, while the remaining four received 5.00, 10.00, 15.00, and 20.00 mL of 15.0 ppm standard Cu^{2+} and were then diluted to the mark. The solutions were measured by flame atomic absorption spectrometry, resulting in respective absorbance values of 0.413, 0.470, 0.534, 0.593, and 0.652. Calculate the concentration of Cu^{2+} present in the lake sample, and determine the 95% confidence interval.

Before plotting the data we need to calculate the final concentration of standard added in each standard addition. For convenience, we will use units of μg of standard Cu^{2+} per milliliter of sample aliquot.

Amt. of standard added (mL)	0	5.00	10.00	15.00	20.00
Conc. standard added (μg standard/mL sample)	0	5.00	10.00	15.00	20.00
Absorbance	0.413	0.470	0.534	0.593	0.652

Using least squares on the data plotted in Figure 16 produces parameter estimates $b_0 = 0.412$ and $b_1 = 0.0120$. The Cu^{2+} concentration for the original 15.00-mL sample aliquot computes to $0.412/0.0120 = 34.3$ ppm. Then the sum of the squares of errors or residuals (SSE) introduced in Section 4.3 will be used to estimate s_y. Hence, $s_y = 1.74 \times 10^{-3}$. For a 95% confidence interval and $5-2 = 3$ degrees of freedom, $t = 3.18$, and the interval equals 34.3 ± 1.3 ppm.

Problem 4. The silver content of a sample was determined by using atomic absorption spectrometry. Five 5.00-mL aliquots were placed in 25.00-mL volumetric flasks, and all solutions were diluted to the mark. Use the following information to estimate the concentration of silver present in the sample, and obtain the 95% confidence interval.

Concentration standard Ag	10.0ppm				
mL standard added	0	5.00	10.00	15.00	20.00
Absorbance	0.214	0.285	0.355	0.420	0.492

An alternative development of the SAM consists of defining our model as

$$r_0 = c_0 k$$

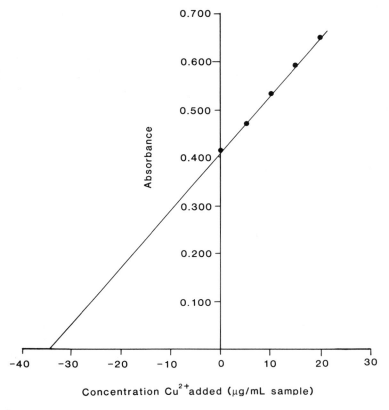

Figure 16 Data plotted for Example 8.

where r_0 identifies the initial response for a sample aliquot that has been diluted to total volume V, c_0 symbolizes the analyte concentration at volume V, and k represents the sensitivity coefficient. After the mth standard addition, the model becomes

$$r_m = c_m k$$

with c_m and r_m representing the total analyte concentration and corresponding response after the mth addition. The change in response with change in concentration is given by

$$\Delta r_m = \Delta c_m\, k$$

CALIBRATION

A least squares fit constrained to pass through the origin can be used to estimate k, the slope. The analyte concentration at volume V predicted by $c_0 = r_0/k$ has a confidence interval of

$$c_0 \pm t\frac{s_{\Delta r}}{k}\left[1 + \frac{r_0^2}{k^2 \Sigma \Delta c^2}\right]^{1/2}$$

for $n-1$ degrees of freedom. The degrees of freedom are reduced by 1 because an intercept term is not estimated.

Example 9. Work Example 8 using regression through the origin. Again, for convenience, we will use units of µg of standard Cu^{2+} per mL of sample aliquot.

mL standard added	5.00	10.00	15.00	20.00
Conc. standard added (Δc) (μg standard/mL sample)	5.00	10.00	15.00	20.00
r_m	0.470	0.534	0.593	0.652
Δr	0.057	0.121	0.180	0.239

Figure 17 shows the data plotted with a fitted line. Estimation of the slope yields $k = 0.120$, giving a predicted sample concentration of $c_0 = r_0/k = 0.413/0.120 = 34.4$. Since the origin is not really a measured data point, the number of degrees of freedom is $4-1 = 3$. Computing $s_{\Delta r} = 2.57 \times 10^{-3}$ and using the above equation, the confidence interval becomes 34.4 ± 1.1 ppm.

Problem 5. Work Problem 4 using regression through the origin.

Multivariate Standard Addition

Saxberg and Kowalski [73] generalized the standard addition (GSAM) to correct for spectral interferences and matrix effects simultaneously. Essentially, the above equations are expanded to vectors and matrices, admitting the inclusion of more analytes and sensors. Key equations are

$$\mathbf{r}_0 = \mathbf{K}'\mathbf{c}_0 \tag{20}$$

and

$$\Delta \mathbf{R} = \Delta \mathbf{C}\mathbf{K} \tag{21}$$

where $\mathbf{r}_0$ identifies the sample response vector with p sensors for sample concentration vector $\mathbf{c}_0$ containing n analytes, $\mathbf{K}$ designates the $n \times p$ calibration matrix, and $\Delta \mathbf{R}$ and $\Delta \mathbf{C}$ are, respectively, changes in responses and concen-

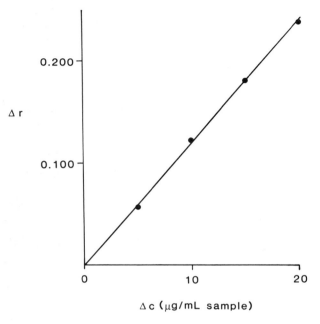

Figure 17 Data plotted for Example 9.

trations for m standard additions. Equation (21) is first solved for **K** using the techniques introduced in Section 4.2. Analyte concentrations are then predicted using Eq. (20).

4.8 SOLUTIONS TO PROBLEMS

1. We are interested not only in determining if the solution is greater than 500 ppm but also in whether it is less than 500 ppm. Hence, the hypotheses are
   $H_0: \mu = 500$ ppm   and $H_1: \mu \neq 500$ ppm
   With $\alpha = 0.05$, $z_{\alpha/2} = 1.96$, $\beta = 0.05$, and $-z_\beta = -1.65$, then
   $$n = \frac{[1.96 - 1.65]^2}{[3]^2/[5]^2} = 36$$

2. The concentration detection limit is $x_d = 3(1.38 \times 10^{-2})/0.0314 = 1.32$ ppm.

3. At the 93.3% level of confidence for the detection limit, $\alpha = \beta = 0.067$. The detection limit can be computed as $y_d = 0.095$. The null hypothesis

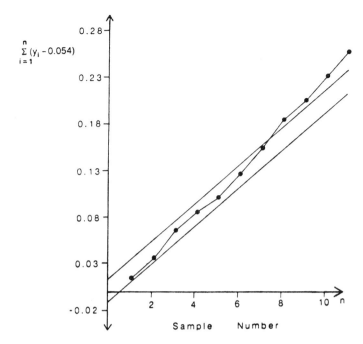

Figure 18 Solution to Problem 3.

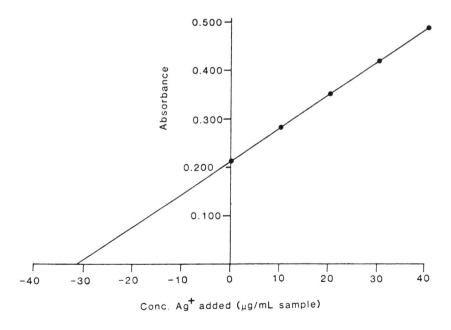

Figure 19 Solution to Problem 4.

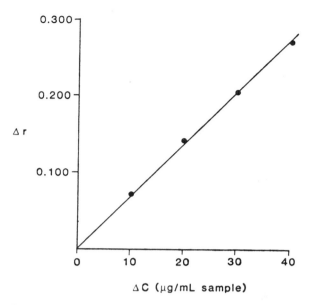

Figure 20 Solution to Problem 5.

is H_0: $y_s = y_b = 0.054$, while the alternative hypothesis is H_1: $y_s \geq y_d = 0.095$. Using sequential sampling (presented in Section 4.1) let $m_0 = 0.054 - 0.054 = 0.0$ and $m_1 = 0.095 - 0.054 = 0.041$. The null hypothesis will be accepted if

$$\sum_{i=1}^{n} [y_i - 0.054] \leq -0.012 + 0.021\,n$$

and H_1 will be accepted if

$$\sum_{i=1}^{n} [y_i - 0.054] \geq 0.012 + 0.021\,n$$

y_i	$\sum_{i=1}^{n} [y_i - 0.054]$	n	y_i	$\sum_{i=1}^{n} [y_i - 0.054]$	n
0.068	0.014	1	0.082	0.153	7
0.075	0.035	2	0.084	0.183	8
0.084	0.065	3	0.074	0.203	9
0.074	0.085	4	0.081	0.230	10
0.069	0.100	5	0.079	0.255	11
0.079	0.125	6			

As Figure 18 shows, after eight measurements the alternative hypothesis is accepted, and Cu is indeed present in the sample.

4. The μg of Ag added per mL of sample are 0, 10, 20, 30, and 40. Figure 19 is a plot of the data yielding $b_0 = 0.215$ and $b_1 = 6.91 \times 10^{-3}$. The Ag concentration present in the original sample equals $0.215/6.91 \times 10^{-3} = 31.1$ ppm. Computing $s_y = 1.89 \times 10^{-3}$, the confidence interval equals 31.1 ± 1.5 ppm.
5. The data are plotted in Figure 20 using $\Delta r = 0.071, 0.141, 0.206$, and 0.278. This results in $k = 6.94 \times 10^{-3}$ and $c_0 = r_0/k = 30.8$ ppm. Calculating $s_{\Delta r} = 5.22 \times 10^{-3}$ gives a confidence interval of 30.8 ± 2.75 ppm.

REFERENCES

1. R. M. Bethea, B. S. Duran, and T. L. Boullion, *Statistical Methods for Engineer and Scientists*, Marcel Dekker, New York. (1985).
2. C. Liteanu and I. Rica, *Statistical Theory and Methodology of Trace Analysis*, Halsted, New York. (1980).
3. J. Cohan, *Statistical Power Analysis for the Behavioral Sciences*, Academic, New York. (1969).
4. A. Wald, *Sequential Analysis*, Wiley, New York. (1947).
5. G. Kateman and F. W. Pijpers, *Quality Control in Analytical Chemistry*, Wiley, New York. (1981).
6. J. M. Hungerford and G. D. Christian, *Anal. Chem.*, 58: 2567 (1986).
7. W. G. Cochran, *Sampling Techniques*, Wiley, New York. (1977).
8. B. Kratochvil and J. K. Taylor, *Anal. Chem.*, 53: 924A (1981).
9. G. Kateman, Chemometrics-sampling Strategies, in *Chemometrics and Species Identification* (Topics in Current Chemistry, Vol. 141), Springer-Verlag, New York. (1987).
10. G. Kateman, *Chemometrics Intell. Lab. Syst.*, 4: 187 (1988).
11. M. A. Sharaf, D. L. Illman, and B. R. Kowalski, *Chemometrics*, Wiley, New York. (1986).
12. H. A. Latinen and W. E. Harris, *Chemical Analysis*, McGraw-Hill, New York. (1975).
13. J. Agterdenbos, F. J. M. J. Maessen, and J. Blake, *Anal. Chim. Acta*, 108: 315 (1979).
14. P. J. Rousseeuw and A. M. Leroy, *Robust Regression and Outlier Detection*, Wiley, New York. (1987).
15. R. Caulcutt and R. Boddy, *Statistics for Analytical Chemists*, Chapman and Hall, New York. (1983).
16. L. L. Juhl and J. H. Kalivas, *Anal. Chim. Acta*, 207: 125 (1988).
17. N. Roberts, J. Sutter, and J. H. Kalivas, *Anal. Chem.*, 61: 2024 (1989).
18. S. D. Frans and J. M. Harris, *Anal. Chem.*, 57: 2680 (1985).

19. H. Mark, *Appl. Spectrosc.*, 42: 1427 (1988).
20. K. Sasaki, S. Kawata, and S. Minami, *Appl. Scpectrosc,*, 40: 185 (1986).
21. F. V. Warren, Jr., B. A. Bidlingmeyer, and M. F. Delaney, *Anal. Chem.*, 59: 1890 (1987).
22. A. Lorber and B. R. Kowalski, *J. Chemometrics*, 2: 67 (1988).
23. M. A. Marus, C. W. Brown, and D. S. Lavery, *Anal. Chem.*, 55: 1694 (1983).
24. I. T. Jolleffe, *Principal Component Analysis*, Springer-Verlag, New York. (1986).
25. T. Naes and H. Martens, *J. Chemometrics*, 2: 155 (1988).
26. C. W. Brown, R. J. Obremski, and P. Anderson, *Appl. Spectrosc.*, 40: 734 (1986).
27. S. M. Donahue, C. W. Brown, and R. J. Obremski, *Appl. Spectrosc.*, 42: 353 (1988).
28. S. M. Donahue, C. W. Brown, B. Caputo, and M. D. Modell, *Anal. Chem.*, 60: 1873 (1988).
29. P. Geladi And B. R. Kowalski, *Anal. Chim. Acta*, 185: 1 (1986).
30. A. Lorber, L. E. Wangen, and B. R. Kowalski, *J. Chemometrics*, 1: 19 (1987).
31. K. R. Beebe and B. R. Kowalski, *Anal. Chem.*, 59: 1007A (1987).
32. D. M. Haaland and E. V. Thomas, *Anal. Chem.*, 60: 1193 (1988).
33. H. Mark, *Anal. Chem.*, 58: 2814 (1986).
34. T. Naes and H. Martens, *Commun. Statist.-Simul. Comput.*, 14: 545 (1985).
35. C. N. Ho, G. D. Christian, and E. R. Davidson, *Anal. Chem.*, 52: 1071 (1980).
36. B. E. Wilson, E. Sanchez, and B. R. Kowalski, *J. Chemometrics*, 3: 493 (1989).
37. H. R. Wilk and S. D. Brown, *Anal. Chim. Acta*, 225: 37 (1989).
38. C. Jochum, P. Jochum, and B. R. Kowalski, *Anal. Chem.*, 53: 85 (1981).
39. J. H. Kalivas, *J. Chemometrics*, 3: 409 (1989).
40. S. Weisberg, *Applied Linear Regression*, Wiley, New York. (1985).
41. D. C. Hoaglin and R. E. Welsch, *Am. Statist.*, 32: 18 (1978).
42. D. A. Belsley, E. Kuh, and R. E. Welsch, *Regression Diagnostics: Identifying Influential Data and Sources of Collinearity*, Wiley, New York. (1980).
43. H. Martens and T. Naes, in *Near-Infrared Technology in the Agriculture and Food Industries* (P. Williams and K. Norris, eds.), American Association of Cereal Chemists, St. Paul, Minn. (1987).
44. F. Critchley, *Biometrika*, 72: 627 (1985).
45. A. Lorber and B. R. Kowalski, *Appl. Spectrosc.*, 42: 1572 (1988).
46. J. H. Kalivas, *Appl. Spectrosc. Rev.*, 25: 229 (1989).
47. R. F. Gunst and R. L. Mason, *Regression Analysis and Its Application: A Data-Orientated Approach*, Marcel Dekker, New York. (1980).
48. D. L. Massart, B. G. M. Vandeginste, S. N. Deming, Y. Michotte, and L. Kaufman, *Chemometrics: A Textbook*, Elsevier, Amsterdam. (1988).
49. A. Lorber and B. R. Kowalski, *J. Chemometrics*, 2: 93 (1988).
50. J. H. Kalivas and P. Lang, *J. Chemometrics*, 3: 443 (1989).
51. A. Lorber and B. R. Kowalski, *J. Chemometrics*, 2: 67 (1988).
52. T. Naes, *J. Chemometrics*, 1: 121 (1987).
53. L. Aarons, *Analyst*, 106: 1249 (1981).
54. G. Puchwein, *Anal. Chem.*, 60: 569 (1988).
55. A. Junker and G. Bergman, *Z. Anal. Chem.*, 278: 273 (1976).

56. H. Mark, *Appl. Spectrosc.*, *42*: 832 (1988).
57. L. A. Currie, *Anal. Chem.*, *40*: 586 (1968).
58. G. L. Long and J. D. Winefordner, *Anal. Chem.*, *55*: 712A (1983).
59. IUPAC, Nomenclature, symbols, units and their usage in spectrochemical analysis—II. Data interpretation, *Anal. Chem.*, *48*: 2294 (1976).
60. H. Kaiser, *Anal. Chem.*, *42*: 26A (1970).
61. P. W. J. M. Boumans, *Spectrochim. Acta*, *33B*: 625 (1978).
62. A. Hubaux and G. Vos, *Anal. Chem.*, *42*: 849 (1970).
63. A. Junker and G. Bergman, *Z. Anal. Chem.*, *272*: 267 (1974).
64. H. Kaiser, *Z Anal. Chem.*, *260*: 252 (1972).
65. A. Lorber, *Anal. Chem.*, *58*: 1167 (1986).
66. D. R. Morgan, *Appl. Spectrosc.*, *31*: 404 (1977).
67. M. Delaney, *Chemometrics Intell. Lab. Syst.*, *3*: 45 (1988).
68. F. C. Garner and G. L. Robertson, *Chemometrics Intell. Lab. Syst.*, *3*: 53 (1988).
69. M. Otto and W. Wegscheider, *Anal. Chim. Acta 180*: 445 (1986).
70. J. H. Kalivas, *Appl. Spectrosc.*, *41*: 1338 (1987).
71. L. L. Juhl and J. H. Kalivas, *Anal. Chim. Acta*, *207*: 125 (1988).
72. J. H. Kalivas, *Talanta*, *34*: 899 (1987).
73. B. E. Saxberg and B. R. Kowalski, *Anal. Chem.*, *51*: 1031 (1979).

5
Nonlinear Regression Analysis

James F. Rusling *University of Connecticut, Storrs, Connecticut*

Many instrumental techniques require some form of mathematical analysis of the raw experimental data to make their results meaningful to the chemist. Mathematical operations may include transforming data into a more desirable form, subtracting background bias, or extracting values of one or more parameters of interest from the data. The latter type of processing may involve analysis of data with mathematical models of expected instrumental response that it is hoped will result in a statistically supported choice of the correct model and reliable estimates of the desired quantities. Linear regression analysis of absorbance versus concentration data on standards is often used in this way to find the molar absorptivity of an analyte and to prepare a calibration line. However, what if the model is nonlinear? For example, suppose we want to find the concentration of analyte (c_a^o) and the dissociation constant (pK_a) for a weak acid from titration data of measured pH versus milliliters of standard base added. Here, the pH is not a linear function of milliliters of base, pK_a, and c_a^o [1]. In such cases, a general numerical approach to least squares analysis called *nonlinear regression* can often be used to find the desired parameters. Like linear regression, the nonlinear version finds the "best fit" of the model to the data. Unlike some other techniques discussed in this book, nonlinear regression requires that well-defined mathematical models be used.

This chapter provides an introduction to the use of nonlinear regression analysis and to the systematic building of models for experimental data. Its aim is to familiarize the reader with the methods, power, and applicability of nonlinear regression analysis for solving chemical problems. The technique is easy to use. For most applications a few simple modifications to general computer programs are all that is needed.

5.1 LINEAR AND NONLINEAR MODELS

The "model" is an equation or set of equations that allows calculation of a dependent variable y, such as an instrumental response, from values of an independent variable x and a set of $k+1$ parameters, $B(i)$. A general mathematical form of such a model for $j = 0$ to n data points is

$$y_j(\text{calc}) = F\{x_j, B(0), B(1), \ldots, B(k)\} \tag{1}$$

Although several independent variables can be used, we limit ourselves here to models with one independent variable.

Regression analysis consists of simultaneous variation of the parameters in the model with respect to a set of experimental data $[y_j(\text{meas}), x_j]$ until the $y_j(\text{calc})$ computed from Eq. (1) are as close as possible to the experimental $y_j(\text{meas})$. This is called "fitting" the model to the data. It is usually done by using the principle of least squares, which holds that the "best" values of the parameters are found when the error sum S,

$$S = \sum_{j=1}^{n} w_j [y_j(\text{meas}) - y_j(\text{calc})]^2 \tag{2}$$

has its minimum value with respect to the parameters. The $y_j(\text{meas})$ in Eq. (2) are experimental data; the $y_j(\text{calc})$ are computed from Eq. (1). The w_j in Eq. (2) are weighting factors; we will discuss their significance later. For now, assume that all $w_j = 1$. This is valid when the random errors in $y_j(\text{meas})$ are independent of the magnitude of $y_j(\text{meas})$.

The simplest model with one independent variable is a straight line:

$$y_j(\text{calc}) = B(0)x_j + B(1) \tag{3}$$

Note that the dependent variable $y_j(\text{calc})$ is a *linear* function of the parameters $B(0)$ and $B(1)$. Thus, Eq. (3) is called a *linear model*. It can be fitted to data by linear regression. In linear least squares analysis, the error sum S is minimized by setting the first derivatives of S with respect to each parameter equal to zero and solving the resulting simultaneous equations for $B(0)$ and $B(1)$. This yields a closed-form equation for each parameter [1], giving the best estimates of slope $B(0)$ and intercept $B(1)$.

NONLINEAR REGRESSION ANALYSIS

Linear regression is possible whenever the dependent variable y is a linear function of the parameters. Applicability of linear regression does not depend on a linear relation between y and x. (From here on, subscripts on y and x will often be dropped for simplicity.) However, experimental data often follow models in which y depends on parameters in a nonlinear way. For example, in

$$y = B(0) + B(1)\log x$$

y is a nonlinear function of x, but it is still linear with respect to the parameters $B(0)$ and $B(1)$, which can be found by linear regression. On the other hand, in the equation describing the kinetics of first-order decay,

$$y = A' \exp[-kt] + B \tag{4}$$

the dependent variable y is a nonlinear function of the rate constant k. Here, t is time, A' is the response at $t=0$, and B represents a constant background signal. If $B=0$, one option is to linearize Eq. (4) by taking the logarithm of both sides:

$$\ln y = -kt + \ln A' \tag{5}$$

Linear regression of the data onto Eq. (5) can now be used, but $\ln y$ is the new dependent variable. The random error in $\ln y$ does not have the same relation to the size of $\ln y$ as does the random error in y to values of y. If random errors in y are independent of y, then errors in $\ln y$ *are not* independent of $\ln y$. The use of $w_j = 1$ is not valid here, and weighted linear regression is required.

The situation is more complex when there is a real background signal, that is, when B is not zero. If B can be measured, $\ln(y-B)$ can be made the dependent variable in a rearranged, linearized version of Eq. (4). If it is not possible to measure B independently, linearization presents a serious dilemma. Nonlinear regression analysis is a fast, easy-to-use alternative to linearization. By employing it, the nonlinear equation (4) can be fit directly to the experimental data by the least squares principle, giving the best values of the parameters A', k, and B. The method is general and is applicable to almost any model. It avoids creating a complex relation between y and its errors, which might result from linearization. Like linear regression, the nonlinear version seeks a minimum in an appropriate error sum, often the quantity S in Eq. (2). Unlike linear regression, closed-form equations for the best values of the parameters cannot be derived when y is a nonlinear function of one or more parameters. Thus, programs for nonlinear regression start from initial "best guesses" of the parameters and use a numerical algorithm to iteratively vary those parameters until the absolute minimum in the error sum is found. This minimum is called the *convergence point* of the analysis and provides the best values of the parameters.

5.2 CONVERGENCE METHODS

General Programs

Reliable general computer programs for nonlinear regression are commercially available. Some of these programs exist in mainframe computer packages, for example, in the Statistical Analysis System (SAS) package. Other versions are designed to run on microcomputers [2,3]. General programs have a "model subroutine" into which the user writes program code for the desired regression model. The model need not be an explicit closed-form equation like Eq. (4). It may be a series of equations or a numerical simulation, as long as it supplies computed values of the dependent variable $y_j(\text{calc})$ to the main program.

With $w_j = 1$, S in Eq. (2) is called an *unweighted error sum*. The approach to minimum S starting from a set of arbitrary initial "best guesses" for k parameters can be viewed as the journey of a point toward the minimum of an error surface in a $(k+1)$-dimensional orthogonal coordinate system. (In practice, the coordinates may not always be fully orthogonal.) One axis of the coordinate system corresponds to S, and the others correspond to the parameters. When $B = 0$ in Eq. (4), a two-parameter nonlinear regression can be done. The error surface has three dimensions. The z axis is associated with the error sum S, while the x and y axes are identified with the parameters A' and k, respectively (Figure 1). From initial guesses of A_i and k_i, an initial point P_i (A'_i, k_i, S_i) is located on the error surface. By systematic variation of the parameters, the algorithm employed by the program to minimize S causes this point to travel toward point P_o (A'_o, k_o, S_o). This is the point of convergence, where S_o is at the absolute minimum on the error surface. The quantities A'_o and k_o are the best values of the parameters, optimized with respect to the experimental data.

General programs for nonlinear regression differ mainly in the algorithm used to minimize S. One commonly used algorithm is the steepest descent method, in which the search for the minimum S travels on the gradient of the error surface. The progress of travel is monitored at each iteration or cycle and is adjusted when necessary. Steepest descent provides fast convergence in the initial stages of the computation but slows down considerably near the convergence point. Such programs written with conservative tolerances for convergence are extremely reliable. Gauss–Newton, another frequently used algorithm, approximates nonlinear models for $y_j(\text{calc})$ by linear Taylor's series expansions [4,5]. After initial guesses, new values of the parameters at each cycle are found by methods similar to linear least squares, using expressions involving first derivatives of S with respect to each parameter. Ideally, each iterative cycle gives successively better estimates for the parameters until the absolute minimum in S is reached. Unlike the steepest descent method, convergence is often fast in the vicinity of the minimum.

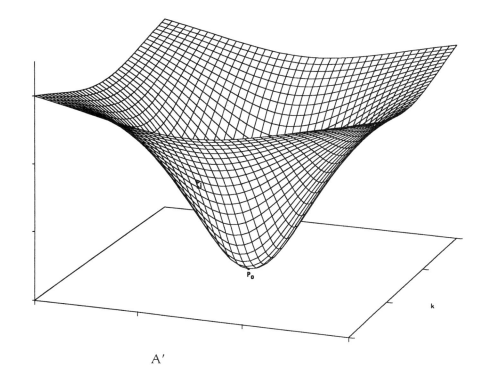

Figure 1 Conceptual representation of the three-dimensional error surface for a two-parameter nonlinear regression analysis. Parameters are A' and k.

The Marquardt or Levenberg–Marquardt algorithm [5,6] contains elements of both the steepest descent and Gauss–Newton methods but converges more rapidly than either of these. The Marquardt algorithm behaves like a steepest descent method under conditions for which the latter is efficient, that is, far from minimum S. Close to the minimum, it behaves like the Gauss–Newton method, again under conditions where the latter is efficient. The Marquardt-based program in reference 3 uses numerical differentiation so that analytical derivatives need not be provided by the user, as required in some Gauss–Newton programs.

Another technique that can be used for nonlinear regression is the modified simplex method. For k parameters, this method involves a $(k+1)$-dimensional geometric construct called a *simplex*. A three-dimensional simplex is a triangle. The minimizing algorithm is such that the simplex simultaneously contracts in size and moves toward the minimum of the error surface [7]. Convergence times are comparable to steepest descent methods but longer than with the Marquardt method.

Nonlinear regression programs can be interfaced with a computer graphics subroutine that plots the experimental data along with the curve computed from the model. A serious mismatch of experimental and computed curves can immediately be seen, and better initial guesses can be made before the regression is started. The graphics subroutine can also be used as a rough visual check of final goodness of fit.

Nonlinear regression programs should include criteria to automatically test for convergence and terminate the program when preset conditions for convergence are reached. Reliable tests for convergence are based on the rate of change of S or on the rate of change of the parameters. For example, one program [2] commonly terminates when the rate of change of S over 10 cycles is <0.02%. Another [3] converges normally when the change in all parameters in a given cycle is smaller than 0.005%. Criteria for convergence can often be adjusted by the user, but this is normally not required. In an algorithm as fast as the Levenberg–Marquardt, testing over a series of cycles is self-defeating in terms of computational time and is usually unnecessary. Supplementary convergence criteria for special situations are provided in some programs.

The job of regression algorithms is to find the absolute or *global* minimum on the error surface. If the surface is irregularly shaped, the algorithm could conceivably hang up in a *local* minimum with an error sum considerably larger than the global minimum. Parameter values would then be in error. Although such false minima are possible, in practice they are rarely encountered in well-conditioned, properly designed regression analyses. The use of computer graphics, mentioned above, also aids in avoiding starting points too far from the absolute minimum that might cause convergence to a false minimum. Local minima can be detected by starting regression analyses with several different (but physically reasonable) sets of initial values of the parameters. Convergence to significantly different values of S and final parameters for different starting points suggests local minima. Convergence to identical values of S and parameters for any reasonable starting point indicates that the global minimum has been found.

Weighting Factors

The relation between random (or standard) errors in dependent and independent variables and the measured values of these variables dictates the choice of weighting factors w_j in the error sum [Eq. (2)] to be minimized. For a model with the general form of Eq. (1), standard errors in both the dependent variable y and the independent variable x may be significant. However, a useful simplifying assumption is that standard error in x is negligible, that is, that the x_j are known with absolute precision. This is a common assumption in linear and nonlinear regression analyses, and we will employ it here. It is reasonably accurate in many cases, such as when y is a slowly varying instrumental response measured versus time as x. Time can often be measured much more accurately and precisely than many instrumental responses. If random errors in x can be neglected, only the random errors in y need be considered in the error sum.

If the standard error in y is independent of the size of y, the error sum S in Eq. (2) can be minimized with all $w_j = 1$. In such cases, we say that there is *absolute* error in y. Suppose we measure the current from an electrolytic cell versus time with a analog strip-chart recorder or by converting the analog signal onto a digital scale of high resolution. As long as the sensitivity scale is the same for all measurements, the standard error in current (y) is likely to be the same whether the current is small or large. Thus, we have *absolute* error in y. This can be checked experimentally by measuring standard deviations of the current, but this is ordinarily not necessary. If the standard deviations are the same throughout the scale used, the errors in y are absolute.

In other cases, standard errors in y may be proportional to y, and the sum of the squares of the *relative* errors should be minimized. This requires setting $w_j = [y_j(\text{meas})]^{-2}$ in Eq. (2). In some cases, standard errors in y may be neither absolute nor strictly relative, and alternative weighting factors are required. This occurs for luminescence decay data obtained by photon counting, for which the random error in y(meas) is Poisson distributed [9], that is, proportional to $[y(\text{meas})]^{1/2}$. Here the proper weighting factor is $w_j = [y_j(\text{meas})]^{-1}$. Any detector that counts single events will require a similar weighting factor. Nonlinear regression programs can be designed with a general error sum as in Eq. (2). Then the user can choose the proper weighting factor before running the program.

Equation (2) involves the sum of the squares of the differences in calculated and experimental values of y. It expresses mathematically the assumption that only the standard errors in y are significant. If random errors in x are also significant, an appropriately weighted function reflecting errors in both x and y should be minimized [8,10].

Example 1. Absorbance is measured versus time for the suspected first-order decay of a chemical reactant. Data are given in Table 1 as YMEAS = A; INDEP. VARIABLE = t in seconds. Assuming absolute errors in A, use non-linear regression to see if first-order decay holds, and find the rate constant for the decay and **A** at **t** = 0.

This example shows how nonlinear regression analysis is used. The data were analyzed by using the program in reference 3. Equation (4) with B = 0 was placed in the model subroutine to analyze these data. In this example y =

Table 1

TEST RUN—FIRST ORDER KINETICS, NORMALLY DISTRIBUTED NOISE ADDED
MATRIX CONDITION NUMBER = 76.18698
CONVERGENCE IN* 6 *CYCLES
STD. DEV. = 3.481841E-03
SUM DELTA SQUARED = 1.454786E-04
PARAMETERS AND ERRORS
B(0) = 5.000978 + ©@ @™㉆®®㉆® −㉆©㉆#
°☒ ★ ☑ = 30
B(1) = .97

(YC-Y)/SD	YCALC	YMEAS	INDEP. VARIABLES
−0.89346	9.048891E-01	9.080000E-01	2.000000E-02
0.50593	8.187615E-01	8.170000E-01	4.000000E-02
−0.33558	7.408316E-01	7.420000E-01	6.000000E-02
2.10204	6.703190E-01	6.630000E-01	8.000000E-02
−1.00009	6.065179E-01	6.100000E-01	9.999999E-02
−0.06051	5.487893E-01	5.490000E-01	1.200000E-01
−0.70212	4.965554E-01	4.990000E-01	1.400000E-01
1.52020	4.492931E-01	4.440000E-01	1.600000E-01
−0.42239	4.065293E-01	4.080000E-01	1.800000E-01
−0.90879	3.678357E-01	3.710000E-01	2.000000E-01
−0.05026	3.328250E-01	3.330000E-01	2.200000E-01
0.32931	3.011466E-01	3.000000E-01	2.400000E-01
0.71325	2.724835E-01	2.700000E-01	2.600000E-01
−0.99133	2.465484E-01	2.500000E-01	2.800000E-01

CORRELATION MATRIX
1.0000E+00   8.0797E-01
8.0797E-01   1.0000E+00
MATRIX PRODUCT
1.0000E+00   0.0000E+00
0.0000E+00   1.0000E+00

absorbance and x = time. We consider time to be free of error. Thus, only random errors in y are significant. Since errors in A are absolute, we minimize S in Eq. (2) with $w_j = 1$. There are two parameters: $B(0) = k$ and $B(1) = A'$. [These data were actually simulated from Eq. (4) with normally distributed noise added, so we know that the true values are $B(0) = 5.00$ and $B(1) = 1.00$.] The regression program (in BASIC) was interfaced with a graphics routine to help choose initial values of parameters. It was run on a PC/XT-type microcomputer. Data can be either input from a disk file or typed into a DATA statement. In response to initial dialogue, the user inputs starting values of the parameters and then views a graph comparing experimental and computed data (Figure 2a). If agreement between the two plots is reasonable, the user begins the regression. In many cases, good initial agreement may require simply that both plots fall in the same coordinate space on the monitor. If agreement is unsatisfactory, a new set of initial parameters can be chosen to give a better graphical comparison.

In our example, we chose a $B(0)$ value rather far from the convergence point, but the Marquardt nonlinear regression analysis converged in six cycles (Table 1). The standard deviation (SD) of the regression of 3.482×10^{-3} is about 0.5% of the largest y value. This is about the same as the amount of absolute random noise added to the simulated absorbances. Values of the computed parameters $B(0)$ and $B(1)$ are in excellent agreement with their true values. Final parameter values are given with their standard errors; small relative values are good indicators of a satisfactory fit and indicate the significance of the parameters in the model. Initial guesses are recorded in the output. Calculated and measured y values and their differences are tabulated for the user's inspection.

The next entry in the printed output (Table 1) is a correlation matrix. Its elements can be interpreted in the same way as the correlation coefficient between y and x variables in a linear regression. The off-diagonal elements of the matrix, elements a_{ij} with $i \neq j$, give an indication of the correlation between the pairs of parameters with the [row,column] index of that element. The only off-diagonal element in our example gives the correlation coefficient between parameters 1 and 2. Diagonal elements of this symmetric matrix should be unity. Correlation is usually not a serious problem if off-diagonal elements are smaller than about 0.98. Two parameters in a regression model that are interdependent are said to be *correlated*. Their final values will depend on one another. If the correlation is total, unique estimates for each parameter will not be found.

The first graphical entry in the printed output is a plot of experimental data and the curve computed from the best values of the parameters (Figure 5.2b). This suggests a good fit, but such plots are insensitive to small deviations from the model. A better discriminator for goodness of fit is the deviation plot (Figure 5.2c), a graph of the normalized deviation at each point, that is, $\Delta y_j = [y_j(\text{meas}) - y_j(\text{calc})]/\text{SD}$ versus x_j. This is sometimes called a *normalized residual* plot. The ordinate is in units of SD. The residual or "deviation" of the jth point is given by

$$\text{dev} = y_j(\text{meas}) - y_j(\text{calc})$$

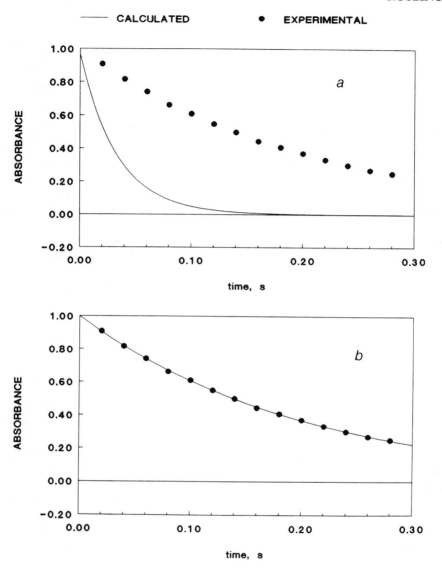

Figure 2 Graphical output for Problem 1. (a) Initial plot before regression analysis; (b) final plot after regression analysis.

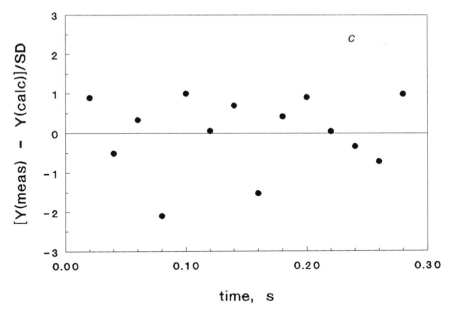

Figure 2c Deviation plot.

Figure 5.2c expresses dev/SD versus time. A deviation plot is a more probing indicator of goodness of fit than summary statistics such as SD because the fit of each data point to the regression model is evaluated separately. A random scatter of points around $\Delta y = 0$ in the deviation plot indicates that the regression model adequately describes the data. Note that points in the deviation plot in the present example are randomly scattered, suggesting a good fit of the model to the data. However, if there are no *systematic* errors in the measurements and the deviation plot is not random but arranged in a pattern, then the model is incorrect. As a supplement to this interpretation, SD should be compared to the estimated standard error (e_y) in measurement of y. The relation SD $\leq e_y$ suggests a rather good fit, and small nonrandom deviations in such instances can often be traced to small systematic errors in measuring y. Systematic errors of measurement can often be uncovered by analysis of data on standard systems known to follow the regression model. Deviation plots in conjunction with summary statistics such as SD are highly recommended for evaluating the quality of fit of the model to the data.

5.3 PRACTICAL MODEL BUILDING

Models for Instrumental Response Curves

We use the term *response curve* to denote a series of digital data points obtained as measured instrumental response versus time or an experimentally controlled variable. Familiar examples are spectral absorbance versus wavelength, current of an electrochemical cell versus potential, and luminescence intensity versus time. The measured response—absorbance, current, or intensity—is the dependent variable y. Wavelength, potential, or time are the respective independent variables x. Regression models for response curves in the form of Eq. (1) should include all contributions to the measured signal, that is, those from nonrandom noise and other instrumental bias as well as the chemical or physical events being studied. A generic expression for such models is

$$y = \text{response of interest} + \text{instrumental background contributions}$$

The terms on the right-hand side may depend on x. Instrumental contributions may include finite background offset, drift, instability, and other characteristics of the measuring system that yield a finite signal. We refer to these instrumental signatures collectively as *background*.

If background contributions are reasonably constant with time or highly predictable, they can sometimes be measured and subtracted from the data. However, if the background can be described in mathematical form, it is often preferable to include specific terms in the model accounting for its contribution. An example of the latter is Eq. (4), where a constant background B is included. If the background contribution drifts linearly with time, B could be replaced with $mt + B$, representing a linear background with slope m and intercept B at $t = 0$.

Closed-Form Theoretical Equations

Theoretical equations for instrumental response versus independent variables are good starting points for regression models. When used with the appropriate background terms, computations are usually fast, and physically significant parameters are obtained. As an example, we discuss response curves of current (i) in an electrolytic cell versus controlled potential (E) varied linearly with time at a working electrode. Under experimental conditions where a steady state exists between the rate of electrolysis and the rate of mass transport to the electrode, sigmoidal i–E curves (Figure 3) result. Such steady-state curves (called *voltammograms*) are found in classical dc polarography, rotating disk voltammetry, and slow-scan voltammetry using disk "ultramicroelectrodes" with radii less than

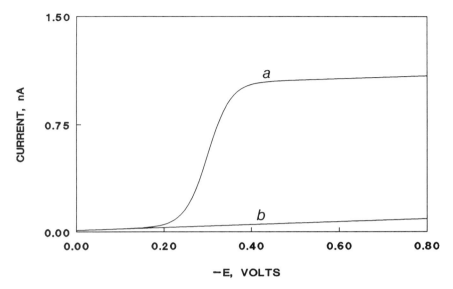

Figure 3 Steady-state microelectrode voltammogram. (a) Curve for reduction of a species in solution; (b) linear background component.

about 10 μm. The simplest electrode process giving rise to a sigmoid steady-state response is one-electron transfer uncomplicated by chemical reactions of reactant (O) or product (R):

$$O + e = R \qquad (E^{0'}) \tag{6}$$

Equation (6) represents a reversible transfer of one electron from an electrode to an electroactive species O. The i–E curve centers around the formal potential $E^{0'}$, which is the same as the half-wave potential ($E_{1/2}$) if the diffusion coefficients of O and R are equal. The half-wave potential is the value of E when the current is one-half the current at the plateau, called the limiting current (i_l). Assuming that only O is present initially in the solution, the equation describing the steady-state faradaic i–E response for Eq. (6) is

$$i = i_l/(1 + \Theta) \tag{7}$$

where $\Theta = \exp[(E-E^{0'})(F/RT)]$, R is the universal gas constant, F is Faraday's constant, and T is the temperature in kelvins.

For a well-resolved i–E curve, an appropriate model [11] for regression analysis combines Eq. (7) with a linearly varying background current:

$$i = \frac{i_1}{1+\Theta} + m(E'-E) + i' \qquad (8)$$

The first term on the right-hand side of Eq. (8) expresses the faradaic current from electron transfer in Eq. (6); the second two terms account for background. E' is an arbitrary potential chosen to fall before the initial current rise so that a background current i' at E' can be measured, and m is the slope of the background current (Figure 3). In nonlinear regression analysis using Eq. (8), i_l, F/RT, and $E^{0'}$ can be used as parameters. E' and i' are correlated, so E' is usually fixed. A correction factor accounting for observed nonparallel plateau and baseline can also be included [11]. The quantities i' and m can be used as regression parameters, or they can be kept fixed. In the latter case, values of i' and m measured from data obtained at potentials before the onset of the wave can be input to the program.

Severely overlapped voltammograms can be separated by using nonlinear regression with a model composed of one term on the right-hand side of Eq. (7) for each of two electroactive components (denoted as 1 and 2):

$$i = i_1 \left(\frac{f}{1+\Theta_1} + \frac{1-f}{1+\Theta_2} \right) \qquad (9)$$

Parameters are $(F/RT)_1$, $(F/RT)_2$, $E_1^{0'}$, $E_2^{0'}$, and f, the fraction of component 1 referred to total analyte in the mixture. Regression onto Eq. (9) successfully resolved overlapped dc polarograms for Pb(II) and Tl(I), which differ in half-wave potential by only about 50 mV [11].

Some species are reduced or oxidized at potentials so close to the ends of the available potential window of the working electrode that their voltammetric waves are severely overlapped with a large increasing background current. Such curves can be resolved by adding a term exponential in E to Eq. (8) [8,11]. The exponential approximately describes the rising portion of the irreversible reaction responsible for the background current at the potential limit of the working electrode system. The model is

$$i = \frac{i_1}{1+\Theta} + m(E'-E) + i' + i' \exp[B'(E'-E)] \qquad (10)$$

where B is an additional parameter proportional to F/RT.

Empirical Models: Peak-Shaped Data

Nonlinear regression can also be used with empirical models if they provide a good quantitative description of the data. For example, single and overlapped peaks in UV–Vis absorbance, fluorescence, X-ray photoelectron (XPS), and

NMR spectroscopy can be reliably fit by a linear combination, of Lorentzian (L) and Gaussian (G) peak shapes, as in the equations

$$y = h[fG + (1-f)L]$$

$$L = \frac{W^2}{(x-x_o)^2 + W^2} \qquad G = \exp\left[\frac{-(x-x_o)^2}{2W^2}\right] \qquad (11)$$

where f is the fraction of Gaussian character, h is peak height, W is half-width at half height, and x_o is the position of the peak maximum. These four parameters are usually used in the regression analysis. Product functions of G and L have also been used [12].

Figure 4 illustrates a fit of simulated data for two overlapping Gaussian peaks. The model used was a linear combination of Eq. (11) with $f = 1$. Thus, the model takes the special form

$$y = mx + y_o + h_1 \exp\left[-\frac{(x-x_{o1})^2}{2W_1^2}\right] + h_2 \exp\left[-\frac{(x-x_{o2})^2}{2W_2^2}\right] \qquad (12)$$

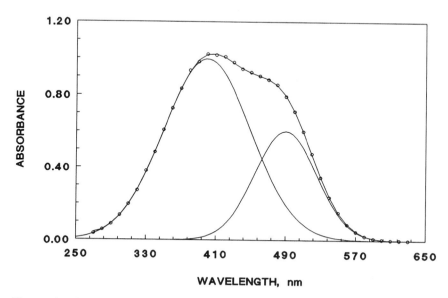

Figure 4 Overapped Gaussian peaks simulated for UV-Vis spectroscopy. Circles are the simulated data points; the line through them was computed from the final parameters of the regression analysis. The reconstructed component peaks are also shown.

where $mx + y_o$ represents a linear background and the last two terms on the right-hand side represent Gaussian peaks 1 and 2. Numerical results were as follows.

Parameter	Value ± error	Actual value	Initial guess
h_1	0.9967 ± 0.0027	0.996	1
x_{o1}	399.8 ± 0.4	399.8	400
W_1	49.99 ± 0.26	49.9	75
h_2	0.6017 ± 0.0065	0.602	0.8
x_{o2}	489.8 ± 0.4	489.8	475
W_2	35.30 ± 0.24	35.3	40

Convergence in eight cycles: SD = 0.0034, S = 0.00036.

Correlation matrix:

$$\begin{matrix} 1 & 0.775 & 0.585 & -0.676 & 0.815 & -0.827 \\ & 1 & 0.912 & -0.944 & 0.944 & -0.844 \\ & & 1 & -0.947 & 0.866 & -0.729 \\ & & & 1 & -0.895 & 0.710 \\ & & & & 1 & -0.835 \\ & & & & & 1 \end{matrix}$$

Parameter values found from the regression analysis are in excellent agreement with their simulation values, and their standard errors are small. The parameter values found by regression can be used to recompute the two contributing peaks (Figure 4). The relative SD of regression is about 0.35% of the largest value of y and smaller than the added "standard error" in y of 0.5%. The deviation plot was random. All of these characteristic "goodness-of-fit" criteria show good agreement between "experiment" and the model.

For real experiments, a simple Gaussian is rarely adequate. From Eq. (11), a Gaussian–Lorentzian model for n' peaks takes the form

$$y = mx + y_o + \sum_{i=1}^{n'} h_i[f_i G_i + (1-f_i)L_i] \qquad (13)$$

and finds more general application. In chromatography, where tailing is often found on one side of the peak, an exponential decay function of x for each peak can be added to Eq. (12) or Eq. (13) [13].

Numerically Solvable Equations

In many cases, models for experimental data cannot be expressed as closed-form equations; no explicit expression of the form of Eq. (1) can be found. This is not an obstacle to applying nonlinear regression provided that a regression program is used that does not require closed-form derivatives of the model. Then a numerical method to compute $y_j(\text{calc})$ at the required values of x_j can be used in the model subroutine of the regression program. One of the simplest cases is when the model takes the form $F(x,y) = 0$. A widely used iterative approximation to find roots of such equations is the Newton–Raphson procedure [4]. If the equation to be solved is of the form $F(y) = 0$ and the initial guess of the root is y_k, then the Newton–Raphson method holds that a better approximation, y_{k+1}, for the root is given by

$$y_{k+1} = y_k - F(y_k)/F'(y_k) \tag{14}$$

where $F(y_k)$ is the function evaluated at y_k and $F'(y_k)$ is the first derivative of the function evaluated at y_k. (The reader is directed to reference 4 for derivations, limitations of the method, and examples of programming.) The method works well for polynomials of degree >2 if a reasonably good initial estimate of the root is provided. Equation (14) is then used repeatedly until convergence to a preset limit (LIM) is reached. A typical convergence criterion is

$$\text{ABS}\left[\frac{y_{k+1} - y_k}{y_{k+1}}\right] \leq \text{LIM} \tag{15}$$

where LIM may be set at 0.0001 or smaller, as desired. Given a good initial estimate, convergence is rapid for well-conditioned applications and is often reached in a few cycles. Equations other than polynomials can also be solved by the Newton–Raphson approach.

Consider a generic example for the use of the Newton–Raphson method within a model subroutine of a nonlinear regression program. Suppose we have the following model for a response curve:

$$y + \exp(x-x_o)y^{2/3} - y_o = 0 \tag{16}$$

where y_o and x_o are constants. We need to use Eq. (14) to find the best root y at each value of x. Here,

$$\begin{aligned}F(y_k) &= y + \exp(x-x_o)y^{2/3} - y_o \\ F'(y_k) &= 1 + (2/3)\exp(x-x_o)y^{-1/3}\end{aligned} \tag{17}$$

In the subroutine used to fit experimental data, Eqs. (14) and (17) are used in combination to find the best root $y_j(\text{calc})$ at each value of x_j. At each cycle of the Newton–Raphson procedure, Eq. (15) is used to test for convergence. Logical statements can be included to terminate any Newton–Raphson routine after 50 or so of its iterations to avoid infinite loops, which could arise during the course of regression analyses.

Testing the Models

In addition to estimating parameters, another major use of regression analysis is to distinguish between several possible models that could give rise to the observed data. The usual goodness-of-fit criteria—deviation plots and summary statistics—are used for this task. Example 1 illustrated the use of SD, standard errors of parameters, and deviation plots for assessing goodness of fit. A small SD with respect to the size of the y_j(meas) along with a random deviation plot can usually be taken as good evidence of an acceptable fit. To distinguish a series of models, the data can be regressed onto each model in succession. If the number of parameters in each model is the same, the best SD consistent with an uncorrelated model and a random deviation plot can be used as the criterion for choosing the best model. The use of summary statistics alone for such comparisons is not recommended because they give little indication of systematic deviations. Deviation plots are often better indicators of goodness of fit than summary statistics because each data point is checked separately for its adherence to the regression model.

If the models being compared have different numbers of regression parameters, a slightly different approach is needed to compare goodness of fit. Here, the individual regression analyses have different numbers of degrees of freedom, defined as the number of data points (n) less the number of parameters (p). The difference in degrees of freedom must be accounted for when using a summary statistic to compare the applicability of two models. A statistical test that can be used in such situations employs the extra sum of squares F-test [8]. This involves calculating the F ratio:

$$F(p_2 - p_1, n - p_2) = \frac{(S_1 - S_2)/(p_2 - p_1)}{S_2/(n - p_2)} \qquad (18)$$

where S_1 and S_2 are the residual error sums [S, Eq. (2)] from regression analyses of the same data onto models 1 and 2, p_1 and p_2 now represent the respective numbers of parameters, and the subscripts refer to the specific models. To use Eq. (18) model 2 should be a generalization of model 1. An example is given in Problem 3 (Section 5.5). Care must be taken in evaluating deviation plots from the models being compared. For fits with about the same SD for models of similar type, an increase in the number of parameters tends to produce greater inherent scatter in the deviation plots. This complicates direct comparisons between closely related models with differences between p_1 and $p_2 \geq 2$.

5.4 APPLICATIONS

Several specific uses of nonlinear regression were discussed above. In this section, a few more of the common applications of nonlinear regression analysis are outlined briefly. More detailed discussions of nonlinear regression analysis in analytical chemistry since the early 1970s are available [1,8], and additional applications are discussed in Section 5.5.

Titrations

Nonlinear regression has been used extensively to analyze acid–base and other titration data by using equilibrium models [1]. Procedures have been directed mainly toward finding the concentration of unknown analyte and equilibrium constants and have been applied to potentiometric, spectrophotometric, and thermometric detection methods. In general, an equation describing the measured response in terms of the initial concentration of analyte (c_a^o), concentration of titrant (c_b), initial volume of analyte (V_a^o), and volume of titrant added (V_b) is used as the model. For example, in potentiometric titrations of a weak monobasic acid with a strong base, the equation [14] for the concentration of hydrogen ion during the titration is

$$[H^+]^3 + \alpha[H^+]^2 - \beta[H^+] - K_a K_w = 0$$

where K_a is the dissociation constant for the weak acid being titrated, K_w is the ion product $[H^+][OH^-]$,

$$\alpha = \frac{c_a^o f}{1+rf} + K_a \qquad \beta = \frac{c_a^o K_a(1-f)}{1+rf} + K_w$$

$$r = c_a^o/c_b$$

The fraction of analyte titrated is $f = V_b c_b / V_a^o c_a^o$. The above third-order polynomial in $[H^+]$ is coupled with

$$pH(calc) = -\log y_{H^+}[H^+]$$

(y_{H^+} is the activity coefficient of hydrogen ion) in the model subroutine to compute pH at each increment of titrant added.

Fitting the pH versus V_b data by nonlinear regression yields highly accurate and precise analyte concentrations and equilibrium constants. Typically, accuracy and precision within 2 ppt can be achieved, even in titrations of weak bases such as acetate, which give poorly defined potentiometric endpoint breaks [1]. One impressive result found by Meites and coworkers was that the concentrations of *both* analyte and titrant could be determined with excellent precision by regression analysis. That is, the titrant need not be standardized precisely! More recent work [15] showed the possibility of detecting a 1% acidic impurity in solutions of an acid with a pK_a differing from that of the analyte by only 0.57 units. Thus, nonlinear regression analysis greatly extends the applicability of classical titration methods. Weighting of the error sum reflecting errors of both measured pH and volume of titrant has been recommended. The reader is referred to the original literature [14] for details.

First-Order Decay Processes

First-order decomposition of chemical species is often found in time-resolved experiments on luminescence and radioactive emission and in kinetics of unimolecular chemical reactions. For mixtures of p components all undergoing first-order decay with lifetimes $\tau_j = 1/k_j$, the data are represented by a sum of exponentials:

$$I = \sum_j A_j \exp\left[\frac{-t}{\tau_j}\right]; \quad j = 1, \ldots, p \tag{19}$$

where I is the measured intensity of the decaying species at a single energy or wavelength and k_j is the first-order rate constant for decay of the jth component. Nonlinear regression analysis of I versus t data can provide the $2p$ parameters A_j and τ_j. If a counting-type detector is used, random errors in the signal are proportional to the square root of the signal. Thus, the weighting factor $w_j = [y_j(\text{meas})]^{-1}$ should be used in Eq. (2). Multichannel decay data collected at a series of energies can be analyzed with an extension of Eq. (19) providing improved precision in parameters [16].

If the number of species undergoing decay is unknown, a judgment must be made concerning the correct number of exponentials in the model. This can be done by an automated computer program that sequentially fits models involving 1, 2, . . . ,p exponentials to the experimental data, stopping when the correct model is found. The choice of the correct model is usually based upon predefined criteria for goodness of fit [8]. Unless decay rates of the components are very different, this method is limited to $p \leq 3$.

Electroanalytical Chemistry

Steady-state voltammetric curves arise in many types of electrochemical experiments, including slow-scan voltammetry at ultramicroelectrodes, rotating disk voltammetry, polarography, voltammetry in flowing streams, and normal pulse voltammetry. The sigmoid i–E curves obtained from these techniques can be analyzed by nonlinear regression by using the models discussed in Section 5.3. The form of the models is the same for all these techniques [8]. Probably the most important analytical use of nonlinear regression in electrochemistry has been for extraction of analytical signals from background and the resolution of overlapping waves from two or more analytes.

Nonlinear regression has also been used for linear sweep voltammetry (LSV), for which general closed-form equations are not available. Digital simulations [17] of the LSVs were used in the model subroutine of a regression program to compute values of current at the required potentials. In this way, rate constants for second-order electron transfer reactions coupled to charge transfer at the electrode were estimated from single LSV response curves. Detailed analyses and discrimination of electrode reaction mechanisms could also be made from the data with this technique [8].

5.5 PROBLEMS AND SOLUTIONS

Problem 1

The data in Table 2 represent first-order luminescence decay of a single species measured by photon counting. The background signal can be considered constant with time. (a) Use unweighted nonlinear regression to find the rate constant for the decay. (b) Use the proper weighting factors to do a second nonlinear regression on the data. Compare results from (a) and (b).

(a) **Unweighted Regression Analysis.** The model is Eq. (4); $w_j = 1$ in Eq. (2). The results were as follows:

Table 2

Signal (counts)	Time (ms)	Signal (counts)	Time (ms)
866	0.1	90	1.6
721	0.2	84	1.7
596	0.3	77	1.8
500	0.4	72	1.9
417	0.5	69	2.0
351	0.6	65	2.1
296	0.7	62	2.2
252	0.8	60	2.3
216	0.9	58	2.4
187	1.0	57	2.5
160	1.1	56	2.6
140	1.2	54	2.7
125	1.3	53	2.8
112	1.4	53	2.9
100	1.5	53	3.0

Parameter	Value ± errors	Initial guess
k	1.9959 ± 0.0028	1
A'	997.04 ± 0.81	1000
B	49.94 ± 0.22	30

Convergence in 7 cycles (deviation plot Fig. 5a): Root mean square (SD) = 0.7118, S = 13.68

Correlation matrix:

```
1   0.6561   0.6952
    1        0.1479
             1
```

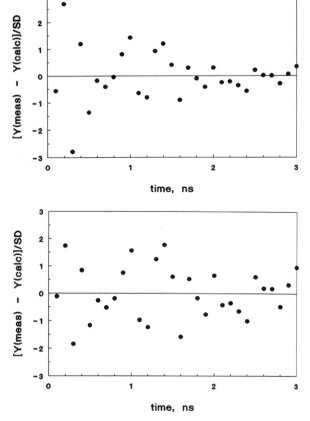

Figure 5 Deviation plots for Problem 1.

NONLINEAR REGRESSION ANALYSIS

(b) Weighted Regression Analysis. $w_j = y_j(\text{meas})$ in Eq. (2). The results are

Parameter	Value ± error	Initial guess
k	1.9944 ± 0.0026	1
A'	997.0 ± 1.2	1000
B	49.88 ± 0.11	30

Convergence in 6 cycles (deviation plot Fig. 5b): Root mean square = 0.0434, S = 0.0509

Correlation matrix:

1	0.7502	0.6807
	1	0.3059
		1

Discussion. Differences in results between weighted and unweighted regressions are not large. Standard errors in the parameters are slightly better for the weighted analysis, and the latter converged in one cycle less. S values are not directly comparable because the w_j are different in the two analyses. A distinct difference appears in the deviation plots, which suggests an uneven distribution of errors in the unweighted analysis.

Problem 2

Microelectrode voltammetry was done at 25°C at a scan rate of 20 mV/s (steady-state conditions) on a 10-μm-radius disk electrode on a solution of an analyte that is reversible reduced in a one-electron reaction. The background current is rather high and variable. Use nonlinear regression of the data in Table 3 onto an appropriate model to find the limiting current and the half-wave potential of the analyte.

The appropriate model is Eq. (8) with $E' = 0$ and parameters i_l, RT/F, $E^{o\prime}$, m, and i'. The results are

Parameter	Value ± error	Initial guess
$E^{o\prime}$	−1.0304 ± 0.0003	−1.025
RT/F	0.02456 ± 0.00048	0.025
i_l	2.17 ± 0.05	2.2
m	−1.03 ± 0.21	−0.8
i'	−0.02 ± 0.19	−0.02

Convergence in 13 cycles: SD = 0.0115, S = 0.00489

Correlation matrix:

1	0.022	0.196	0.341	0.340
	1	0.916	0.820	0.810
		1	0.970	0.967
			1	0.999
				1

Table 3

$-E$ (V)	i (nA)	$-E$ (V)	i (nA)
0.900	0.9058	1.015	1.758
0.905	0.9130	1.020	1.887
0.910	0.9270	1.025	2.015
0.915	0.9387	1.030	2.113
0.920	0.9549	1.035	2.227
0.925	0.9688	1.040	2.338
0.930	0.9827	1.045	2.466
0.935	0.9849	1.050	2.548
0.940	1.0107	1.055	2.660
0.945	1.0246	1.060	2.814
0.950	1.0497	1.065	2.874
0.955	1.0646	1.070	2.960
0.960	1.052	1.075	2.960
0.965	1.096	1.080	3.011
0.970	1.134	1.085	3.065
0.975	1.190	1.090	3.104
0.980	1.219	1.095	3.112
0.985	1.279	1.100	3.167
0.990	1.351	1.105	3.186
0.995	1.432		
1.000	1.496		
1.005	1.589		
1.010	1.699		

Parameters $E_{1/2} = E^{o\prime}$ and i_l are given directly. RT/F is close to its theoretical value of 0.02569 V^{-1}. Background parameters m and i' are strongly correlated, but this is not important because their exact values are not needed.

Problem 3

Diffusion in a solution of macromolecules was studied by measuring apparent diffusion coefficients (D') at a series of concentrations of a small probe molecule (X) that binds to macromolecules (M). Assume a rapid equilibrium:

$$M + nX \rightleftharpoons MX_n \qquad K^n = \frac{[MX_n]}{[M][X]^n}$$

where K is the apparent equilibrium constant per bound X. Two such equilibria will be present if there are two macromolecules M_1 and M_2. Analyze the two sets of data given in Table 4, and find (a) if the solution contains one macromolecule

Table 4

	C_X (mM)	$10^6 \, D'$ (cm^2/s)
Set 1	0.10	4.82
	0.20	2.34
	0.40	1.56
	0.60	1.35
	0.80	1.33
	1.00	1.26
	2.00	0.878
	3.00	0.875
	4.00	0.854
	5.00	0.873
Set 2	0.20	5.60
	0.40	4.07
	0.60	3.13
	0.80	2.02
	1.00	1.89
	1.50	1.48
	2.00	1.23
	3.00	0.93
	4.00	0.77
	5.00	0.83

Note: Total conc. of macromolecules is construct.

or two macromolecules of different sizes (i.e., different diffusion properties) and (b) the diffusion coefficients of the macromolecules.

The apparent diffusion coefficient measured for the system (D') depends on the properties of X and M and can be expressed [18] for one macromolecule as

$$D' = f_a D_o + f_b D_1 \tag{20}$$

where D_o is the diffusion coefficient of X, D_1 is the diffusion coefficient of M, and f_a and f_b are the fractions of unbound and bound X. Since the sum of the fractions must equal unity, $f_a = 1 - f_b$. For two macromolecules,

$$D' = f_a D_o + f_{b1} D_1 + f_{b2} D_2 \tag{21}$$

where $f_{b1} D_1$ refers to M_1 and $f_{b2} D_2$ to M_2. Models for D' can be constructed by expressing the fractions of free and bound probe in terms of the equilibria involved. It is useful to relate D' to the total concentrations of X (C_X) and M (C_M). Using the symbol $x = [X]$, we have [18]

$$C_X = x + [M] n K'^n x^n \tag{22}$$

$$C_M = [M] + [M]K^n x^n \qquad (23)$$

$$f_a = \frac{x}{C_X} = \frac{1}{1 + [M]nK^n x^{n-1}} \qquad (24)$$

$$f_b = \frac{[M]nK^n x^{n-1}}{1 + [M]nK^n x^{n-1}} \qquad (25)$$

Substituting Eqs. (24) and (25) into Eqs. (20) and (21) yields the two models for D' expressed as functions of K, $[M]$, and x. Simplifying assumptions and approximations are made such that the known total concentrations of M and X appear in the models rather that $[M]$ and x. If $x \ll C_X$, that is, if the probe is tightly bound, we have the approximate equation

$$D' = \frac{D_0}{1 + C_M K^n C_X^{n-1}} + \frac{D_1 C_M K^n C_X^{n-1}}{1 + C_M K^n C_X^{n-1}} \qquad (26)$$

for the single macromolecule, where K is an apparent binding constant per bound X. Using the same approach for two macromolecules yields

$$D' = \frac{D_0}{1 + C_{M1} K_1^n C_X^{n-1} + C_{M2} K_2^m C_X^{m-1}} + \frac{D_1 C_{M1} K_1^n C_X^{n-1}}{1 + C_{M1} K_1^n C_X^{n-1} + C_{M2} K_2^m C_X^{m-1}} + \frac{D_2 C_{M2} K_2^m C_X^{m-1}}{1 + C_{M1} K_1^n C_X^{n-1} + C_{M2} K_2^m C_X^{m-1}} \qquad (27)$$

where C_{M1} and C_{M2} are the total concentrations of M_1 and M_2, respectively. Equation (26) was used with parameters D_0, D_1, and $C_M K^n$. Equation (27) was used with parameters D_0, D_1, D_2, $C_{M1} K_1^n$, and $C_{M2} K_2^m$.

In this problem the exponents n and m were fixed at integer values but varied by one unit in a series of sequential nonlinear regression analyses until the smallest values for SD were reached. For example, when using Eq. (26), sequential regressions were done with $n = 1,2,3,4,5$. The small number of data, their relatively large random noise (about $\pm 10\%$), and the different numbers of parameters in the two fits make the deviation plots less useful for discriminating between models in this problem.

Data Set 1. Using Eq. (26) for data set 1, the smallest SD and a random deviation plot were found for $n = 3$, giving the following results.

Parameter	Value ± error	Initial guess
D_0	9.21 ± 2.1	7
D_1	1.01 ± 0.07	2
$KC_M^{1/n}$	4.88 ± 0.74 ($n=3$)	3

Convergence in 7 cycles: SD = 0.177, S = 0.219

Correlation matrix:

$$\begin{matrix} 1 & 0.371 & 0.985 \\ & 1 & 0.430 \\ & & 1 \end{matrix}$$

For the five-parameter fit to Eq. (27), a large number of regression analyses with n and m at whole-number values >2 showed that $n=4$, $m=8$ gave the best SD and a random deviation plot. The results were as follows.

Parameter	Value ± error	Initial guess
D_0	6.84 ± 0.24	7
D_1	1.39 ± 0.03	1.5
D_2	0.85 ± 0.02	0.8
$K_1C_{M1}^{1/n}$	4.93 ± 0.14 ($n=4$)	5
$K_2C_{M2}^{1/m}$	2.00 ± 0.11 ($m=8$)	3

Convergence in 12 cycles: SD = 0.0372, S = 0.00692

Correlation matrix:

$$\begin{matrix} 1 & 0.526 & 0.931 & 0.965 & 0.554 \\ & 1 & 0.194 & 0.632 & 0.789 \\ & & 1 & 0.113 & 0.385 \\ & & & 1 & 0.629 \\ & & & & 1 \end{matrix}$$

The five-parameter regression gives a much lower standard deviation and much lower standard errors of the parameters. Both deviation plots were random. To distinguish between the two models we must use the F ratio in Eq. (18). Thus,

$$F(2,5) = \frac{(0.219 - 0.0069)/2}{0.0069/5} = 77$$

From F tables, $F(2,5)$ at the 99% confidence level is 18.31, much less than the experimental value. This shows that the five-parameter model gives the best fit at the 99% level of confidence.

Data set 2. Using Eq. (26) the results were as follows.

Parameter	Value ± error	Initial guess
D_0	6.44 ± 0.23	6
D_1	0.81 ± 0.08	2
$KC_M^{1/n}$	1.64 ± 0.07 ($n=3$)	4

Convergence in 6 cycles: SD = 0.1445, S = 0.1462

Using Eq. (27):

Parameter	Value ± error	Initial guess
D_0	5.87 ± 0.19	7
D_1	1.43 ± 0.18	1.6
D_2	0.79 ± 0.11	0.8
$K_1 C_{M1}^{1/n}$	1.74 ± 0.09 ($n=4$)	1.7
$K_2 C_{M2}^{1/m}$	0.82 ± 0.12 ($m=9$)	1

Convergence in 6 cycles: SD = 0.1478, S = 0.1094

Here, the five- and three-parameter regressions give similar standard deviations. Standard errors of the parameters do not improve in the five-parameter fit. Again, both deviation plots were random. The F ratio in Eq. (18) is

$$F(2,5) = \frac{(0.146 - 0.109)/2}{0.109/5} = 0.85$$

$F(2,5)$ at the 80% confidence level is 3.78, much larger than the above value. Thus, there is little confidence in the five-parameter model, and the three-parameter model gives the best fit.

REFERENCES

1. L. Meites, *CRC Crit. Rev. Anal. Chem.*, 8: 1 (1979).
2. L. Meites, *The General Nonlinear Regression Program CFT4A*, privately published, Potsdam, N.Y. (1983). (L. Meites, Department of Chemistry, George Mason Univ., Fairfax, VA 22030.)
3. CET Research Group, *Nonlinear Least Squares Program NLLSQ*, privately published (P. O. Box 2029, Norman, OK 73070.)
4. J. K. Johnson, *Numerical Methods in Chemistry*, Marcel Dekker, New York, 1980.
5. S. D. Christian and E. E. Tucker, *Am. Lab.* 31–36, (Sept. 1982).
6. D. W. Marquardt, *J. Soc. Ind. Appl. Math.*, 11: 431 (1963).
7. S. N. Deming, and S. L. Morgan, *Anal. Chem.*, 45: 278A (1973).
8. J. F. Rusling, *CRC Crit. Rev. Anal. Chem.* 21: 49 (1989).

9. J. N. Demas, *Excited State Lifetime Measurements*, Academic, New York. (1983).
10. Y. Bard, *Nonlinear Parameter Estimation*, Academic, New York. (1974).
11. L. Meites and L. Lampugnani, *Anal. Chem.*, 45: 1317 (1973).
12. P. M. A. Sherwood, in *Practical Surface Analysis* (D. Briggs and M. P. Seah, eds.), Wiley, New York, pp. 445–475. (1983).
13. F. J. Knorr, H. R. Thorsheim, and J. M. Harris, *Anal. Chem.*, 53: 821 (1981).
14. G. Kateman, H. C. Smit, and L. Meites, *Anal. Chem. Acta*, 152: 61 (1983).
15. L. Meites, *Anal. Lett.*, 15(A5): 507 (1982).
16. F. J. Knorr and J. M. Harris, *Anal. Chem.*, 53: 272 (1981).
17. J. A. Arena and J. F. Rusling, *Anal. Chem.*, 58: 1481 (1986).
18. J. F. Rusling, C.-N. Shi, and T. F. Kumosinski, *Anal. Chem.*, 60: 1260 (1988).

6
Experimental Design and Optimization

M. J. Adams *Wolverhamptom Polytechnic, Wolverhampton, England*

Experimentation provides the practical foundation on which our knowledge of the world and environment is based. It is the purpose of any experiment to provide information and so answer questions. A good experimental design will enable the answers to be obtained with the minimum effort. The aim of this chapter is to provide the reader with the necessary background to appreciate those aspects of chemometrics applicable to the design and efficient exploitation of experiments.

In general, an experiment is used to investigate the effect, as some suitable response measure, of changing one or more of the conditions defining the experimental system. An everyday practical example will serve to introduce the statistical terms commonly employed in experimental design and data analysis. Consider the making of a cup of tea. In an attempt to appreciate the influence of such *factors* as water temperature, quantity of selected tea, infusion time, and volume of added milk, an experiment can be designed. The factors, or *variables*, define the boundaries of the experiment, with each assuming a range of *levels* against which we measure the taste of the tea, our response variable. By investigating the *treatments*, the combinations of factors and levels, our experiment can provide us with the information required, whether this is a detailed statistical study of the effect of each variable or some overall conclusion as to the conditions necessary for making the "best" cup of tea. Of course, it may be the case that in addition to the controlled factors, set at selected levels, there exist *uncontrolled factors*. Although the state of an uncontrolled variable may not be set by the experimenter, it is usually assumed, by careful selection of experimental conditions, to have a random influence on the experimental results. Its effect will average out to zero over the whole experiment. If, for example, time of day is thought to influence the taste of our tea, then our experiments should be repeated at different times on different days to randomize this uncontrolled effect.

A successful experiment has several prerequisites, the most important of which are (1) clearly stated objectives and (2) an estimate of the acceptable level of experimental error in the result.

To obtain suitable conclusions from any experiment, it is necessary to identify all the factors that can affect the result and include them in the experimental design plan, with the aim of minimizing the effects of the uncontrolled factors. For the preliminary design plan the experimenter must rely on prior knowledge of the system, past experience, and intuition in selecting the relevant factors for study. Having chosen the factors, the allowable or acceptable levels of each must be determined. In some cases these values are immediately apparent—for example, simply *on* or *off* for a switch. In other cases, the levels, and their resolution, must be determined with care to enable the experiment to be described with the fewest possible treatments while providing the detail of data required from the experiment.

When the best design has been produced and the practical study undertaken, then the results from the experiment can be subjected to statistical analysis, and tests on the significance of individual factors and their effects can be made. A number of underlying assumptions are taken in using the more common statistical tests.

First, the uncontrolled experimental errors should be independent and random. This is particularly important when applying the t test and F test. Trends in experimental conditions due to uncontrolled factors can produce nonrandom, dependent errors. The effect can largely be overcome by randomizing the tests, and as a basic precaution randomization should always be followed. A satisfactory experimental design requires that the data produced should be free of bias, which in practice means that any effect, or combination of effects, not under control should contribute randomly to the experiment and their average contribution should be zero. It is crucial to the successful outcome of the experiment to recognize the major sources of nonrandom, systematic error and to allocate the experimental tests in a random fashion to the various combinations of effects from these sources. Where randomization is undertaken, tables of random numbers, or some alternative device, should be used to assign tests.

It frequently arises that known sources of systematic error can be identified but cannot be controlled economically. In such cases the experiment may be designed so that the required factor effects and comparisons are independent of the effects of the principal sources of error. A technique commonly employed to implement these types of experiments involves *blocking*, a procedure that limits the effects of variability in experimental conditions. For example, when comparing analytical methods for accuracy and precision, it would be most inadvisable to examine one method on day 1, a second method on day 2, etc. The day-to-day variation could subject the experiment to many uncontrolled factors that might swamp the observed variation between methods. It is intuitively better to group—that is, block—the tests so that each method is examined, in a random sequence, on each day. Such a randomized block is the simplest of the experimental arrangements available for improving the precision of comparisons.

Second, it is usual to assume that the variance of errors within groups of measurements is similar. In practice, if the number of observations in each group is the same, this criterion may not be too important. Transformation methods, such as taking logarithms, can be used to modify data if necessary. The experimental errors should be normally distributed, although the most commonly employed statistical tests are sufficiently robust and insensitive to minor deviations from theoretical normality. In practice, the likely departure from normality observed in most laboratory work has little effect on the use of the F test and t test.

Most experiments can be assigned to one of two major classes, comparative or factorial. In *comparative experiments* the object of the experiment is to compare the effects of different treatments on the observed response. The results provide data from which we may infer how individual factors influence the chosen response. The aim of *factorial experiments* is to investigate how the result or response is affected by simultaneous changes in the different factors and, commonly, to find the combination of factor levels producing the maximum, or minimum, value for the response variable, that is, *optimization*.

While the calculation of statistical values is relatively trivial, the operations are often repetitive and tedious. The use of tabulated methods aids in preventing errors, and use can be made of spreadsheet software available on all modern computer systems. As well as aiding in the analysis of data, a spreadsheet can be invaluable in studying the analytical procedures. In particular, the "what if" capability afforded by the automatic recalculate facility following the changing of data enables the nature of the study to be appreciated. Tabulated schemes are used extensively throughout this section, and the reader is encouraged to use a spreadsheet to confirm the results shown and to experiment with the techniques. Similarly, the statistics texts by Youmans [1] and Brookes et al. [2] provide numerous examples of tables that can be adapted for use with spreadsheets.

6.1 RANDOMIZED BLOCKS

Randomization is a simple yet effective means of eliminating the effects of systematic errors in many experiments. It may not always be possible to fully randomize an experiment, however. Consider the case of analyzing a set of water samples over a period of several days. Observations and measurements made on any day by one technician will generally show better agreement than those made on different days by different staff members. Such between-staff and between-day effects can be minimized by dividing the tests into blocks and ensuring that the order of tests within any block is randomized. Comparisons can then be made on data produced within a block, eliminating between-block errors.

In all but the simplest experiments, more than two factors or treatment means are to be compared. In such cases, analysis of variance (ANOVA) can used to separate the variation caused by changes in the controlled factor from the variation due to random experimental error.

Consider a laboratory experiment to compare three independent methods of analysis proposed for the determination of the concentration of nickel in a steel alloy. A sample of the alloy can be weighed and dissolved in acid to produce a standard stock solution for subsequent analysis. For the three analytical methods examined, the results can be recorded as in Table 1.

ANOVA provides a scheme whereby the total observed variation is partitioned into two components, one giving the variability between group means and the other the within-group variation. The variance values calculated can be ratioed and compared using the F test. The results for the above analysis can be tabulated as follows.

Source of variation	Degrees of freedom	Sum of squares	Mean sum of squares	F ratio
Between methods	2	0.143	0.0715	11.12
Within methods	21	0.135	0.0064	
Total	23	0.278		

From statistical tables, $F_{(0.01,2,21)} = 5.78$. Thus, the result of 11.12 is significant at the 1% level, and the conclusion is that there is a real, significant difference between the three methods.

It may well be the case that the experiment described above to compare different analytical methods is complicated further by the work being undertaken by different staff members. If two analysts are used in the experiment, it is obviously necessary for each analyst to use all three methods on a random

Table 1 Concentration of Nickel (mg/kg) in Steel by Three Methods

Sample no.	Analytical method		
	I	II	III
1	2.82	2.91	3.12
2	2.81	2.93	2.83
3	2.79	2.98	2.84
4	2.80	2.99	3.02
5	2.81	3.02	2.84
6	2.80	3.12	2.84
7	2.85	3.04	2.78
8	2.74	2.94	2.95
Mean	2.80	2.99	2.90
Sums	22.42	23.93	23.22
Square of sums	502.656	572.645	539.168

Sum of all values = 69.57
Correction factor (CF) = $69.57^2/24$ = 201.666
Sum of all data squares = 201.944
Corrected sum of squares = 201.944 − 201.666 = 0.278
Between-groups (i.e., methods) sum of squares = [(502.656 + 572.645 + 539.168)/8] − 201.666 = 0.143

selection of samples, that is, to use a randomized block design. In such a case the resultant data may be tabulated as in Table 2.

The interaction between analyst and method can be estimated by subtracting the between-methods sum of squares and the between-analyst sum of squares from the between-blocks sum of squares,

Interaction sum of squares = 0.161 − (0.143 + 0.016) = 0.002

Tabulating the results gives

Source of variation	Degrees of freedom	Sum of squares	Mean squares	F ratio
Between methods	2	0.143	0.0715	11.00
Between analysts	1	0.016	0.0160	2.46
Interaction	2	0.002	0.0010	0.15
Experimental (random)	18	0.117	0.0065	
Totals	23	0.278		

Table 2 Concentration of Nickel (mg/kg) Using Three Methods and Two Analysts

Analyst	Sample no.	Analytical method			Total	Total squared
		I	II	III		
A	2	2.81	2.93	2.83		
	3	2.79	2.98	2.84		
	5	2.81	3.02	2.84		
	8	2.74	2.94	2.95		
Sum (A)		(11.15)	(11.87)	(11.46)	(34.48)	(1188.870)
Squares of sums		(124.323)	(140.897)	(131.332)		
B	1	2.82	2.91	3.12		
	4	2.80	2.99	3.02		
	6	2.80	3.12	2.84		
	7	2.85	3.04	2.78		
Sum (B)		(11.27)	(12.06)	(11.76)	(35.09)	(1231.308)
Squares of sums		(127.013)	(145.444)	(138.298)		
Sums (A + B)		22.42	23.93	23.22		
Sum of squares		502.656	572.645	539.168		

Sum of all values = 69.57
Correction factor = 201.666
Sum of all squares = 201.944
Corrected sum of squares = 201.944 − 201.666 = 0.278
Between-methods sum of squares
 = [(502.656 + 572.645 + 539.168)/8] − 201.666 = 0.143
Between-analysts sum of squares
 = [(1188.870 + 1231.308)/12] − 201.666 = 0.016
Between-blocks sum of squares
 = [(124.323 + 140.897 + 131.332 + 127.013 + 145.444 + 138.298)/4]
  − 201.666
 = 0.161

From statistical tables, $F_{(0.05,2,18)} = 3.55$ and $F_{(0.05,1,18)} = 4.41$. Therefore, we can conclude that the methods give significantly different results. There is no evidence, at the 5% level, of there being bias due to analyst A or B, and, similarly, there is no indication of analyst–method interaction. The difference between the methods is independent of the analyst conducting the tests.

6.2 LATIN SQUARES

Although the randomized block design of experiments probably forms one of the most important and most common types of comparative experimental plan, a modification referred to as Latin square designs can prove useful. By using a carefully balanced design strategy it may be possible to eliminate the effects due to more than one blocking variable. When producing a blocked design of experimental trials it may be possible to propose two different ways of assigning the tests to blocks. For example, in the examples discussed above, if the methods were compared using different analysts working over several days, should the results be blocked according to the analyst or according to the day of the trial? For such cases and where there are equal numbers of blocks and treatments and there is known to be no interaction between factors, a Latin square design can be implemented.

This plan is illustrated in the next example. The choice of analytical method for the determination of lead adsorbed on three types of glass fiber filters is to be examined. If three independent methods are available and three people can perform the analyses, then sets of three samples as similar as possible are required to test each method with each staff member. In this example, if three solutions are prepared from each sample of filter adsorbate, which factor should form the blocking element? If it is believed that there is little difference between the samples, then the choice of analysts for the blocks is obvious. If there is confidence that the analysts can provide essentially identical results, then the filter samples will form the blocks. Fortunately, with the aid of a Latin square design, this choice does not have to be made. The design makes use of a tabulated experimental plan in which each treatment appears exactly once in each row and each column of the table.

To design our experiment, we can consider the nine lead solutions, identified by lowercase letters (a, . . . ,i), as being arranged in a square.

Analyst	Filter number		
	1	2	3
A	a	b	c
B	d	e	f
C	g	h	i

If all the solutions were analyzed using a single method, then a two-factor analysis of variance would reveal the significance of any different between filter samples and analysts. As we wish to investigate the effect of choice of method, the assignment of samples to the methods is restricted. In practice, the three solutions from each filter are assigned at random to the three methods, subject to the restriction that no filter or analyst is represented more than once with each method. If we designate the three analytical methods under study as methods I, II, and III, then one arrangement that satisfies the imposed criteria is

	Filter number		
Analyst	1	2	3
A	a(I)	b(II)	c(III)
B	d(III)	e(I)	f(II)
C	g(II)	h(III)	i(I)

This is one arrangement randomly selected from the 12 possible designs. The sets of samples assigned to methods have been balanced with respect to both filter samples and analysts. Examining the actual analytical data obtained from such an experiment (Table 3) will illustrate the use of a Latin square design and, proceeding in the usual manner, subsequent data processing.

Table 3 Mass of Lead (µg) per Filter

Analyst	Filter no.			Total
	1	2	3	
A	40 (I)	52 (II)	50 (III)	142
B	52 (III)	38 (I)	54 (II)	144
C	42 (II)	45 (III)	32 (I)	119
Total	134	135	136	405
	I	II	III	
Method totals =	110	148	147	

Correction factor = $405^2/9$ = 18,225
Corrected sum of squares
 = $(40^2 + 52^2 + \ldots + 45^2 + 32^2)/3 - 18{,}225 = 456$

Between-filters sum of squares
 = $(134^2 + 135^2 + 136^2)/3 - 18{,}225 = 0.667$
Between=analysts sum of squares
 = $(142^2 + 144^2 + 119^2)/3 - 18{,}225 = 128.667$
Between-methods sum of squares
 = $(110^2 + 148^2 + 147^2)/3 - 18{,}255 = 312.667$

Tabulating the results, we have

Source of variation	Degrees of freedom	Sum of squares	Mean squares	F Ratio
Between filters	2	0.667	0.333	0.048
Between analysts	2	128.667	64.333	9.190
Between methods	2	312.667	156.333	22.333
Experimental	2	14.000	7.000	
Totals	8	456.000		

From tables, $F_{(0.05, 2, 2)} = 19.00$, and thus the difference between the methods is highly significant and worthy of further study. The difference between analysts should be considered in future experiments. There is no significant difference between the effects of types of filters on the the analytical procedures.

In chemical applications it is not common for this procedure to be extended to cover more factors, because the assumption of no significant interaction between the variables can rarely be made. The simple Latin square design can be invaluable for an initial investigation in which trials are made to check for large changes. In these preliminary designs, any interaction effects are likely to be small compared with the main single-factor effects.

6.3 FACTORIAL DESIGNS

A factorial experimental design is a plan that aims to test all permitted combinations of all allowed levels of each variable, enabling the major effects and interactions between factors to be estimated. The conventional, so-called classical, investigative scheme chooses each factor or variable in turn for study at a number of discrete selected levels while holding constant the values of each remaining variable. For an experiment designed to investigate the influence of four factors, each studied at three levels, then $4(3-1)+1$, or 9, experimental combinations of variables must be examined in a classical one-variable-at-a-time approach. In comparison, a complete factorial experiment will involve, for this same example, 3^4, or 81, treatments. The factorial design provides a complete picture of the system under study; it is comprehensive and of wider application because the trials undertaken cover a greater range of experimental conditions. Despite the increased cost, in both time and materials, associated with the factorial design compared with the classical approach, it is of particular importance in illustrating and highlighting the presence of interactions between the experimental factors. For this reason, in terms of the information produced by a given amount of experimental effort, a factorial plan can provide for the most efficient experimental design.

Interaction

The major advantage of the complete factorial experiment compared with the single-variable-at-a-time method is that it is sensitive to interactions between factors. *Interaction* is a statistical term used to describe an observed bias in experimental data arising from nonadditive effects of the treatments studied. The relevance of this to experimental design can be best illustrated with an example. If a chemist is interested in determining the absorbance of an analytical solution as a function of solution pH and temperature, an initial study could involve examining the effect of each factor at several levels. Using the one-at-a-time procedure, the results obtained could be as illustrated in Figure 1.

From these data, then, investigating the effect of pH at constant temperature leads to an optimum value of pH 6. At this pH the optimum reaction temperature is 25°C. The conclusion from these trials indicates a maximum absorbance response of 0.5 at pH 6 and 25°C. The results indicated in Figure 1 show that if either temperature or pH is individually varied from these conditions, a decrease in response is observed. The results, however, do not provide any information as to the likely effects of changing the treatments simultaneously—the effects due to an interaction between pH and temperature. The real situation may be as shown in Figure 2, with a true optimum response of about 0.8 absorbance unit at pH 7 and a temperature of 30°C. The one-at-a-time approach fails to provide complete information about the system because it assumes independent, additive effects from each factor, which is often not the case. To obtain the overall picture of the system—not only the main individual factor effects but also their mutual interactions—it is necessary to perform a complete factorial experiment.

Clearly, as the numbers of factors and their possible and permitted levels increase, the number of potential trials in a complete factorial design can quickly become practically prohibitive. To overcome this problem, limited, special-case plans are commonly employed. A useful design procedure for preliminary trials in those cases involving a large number of variables is the two-level factorial design. With n factors this is referred to as a 2^n *factorial design*. Such a limited design is important not only because it employs relatively few trials while covering a wide range of factor space but also because the results can indicate major trends and effects as well as interactions between variables.

2^n Factorial Designs

A particular notation is used to record the data from 2^n factorial designs. The use of this scheme can best be illustrated by reference to a simple three-factor experiment involving the variables A, B, and C, each of which will be present at only two levels, high and low. The choice of the actual levels selected for each variable in the experiment is largely a technical matter based on experience

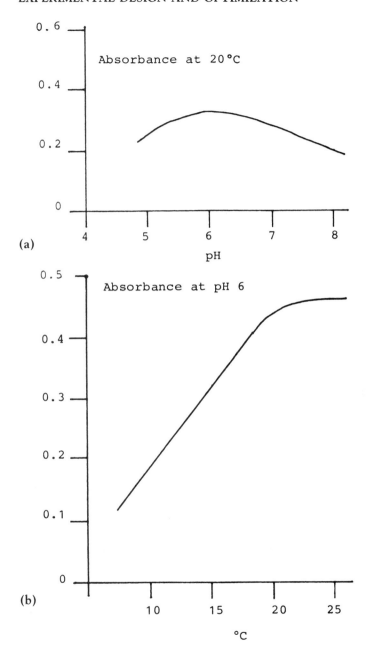

Figure 1 The effect of (a) pH and (b) temperature studied one at a time on the analytical solution absorbance.

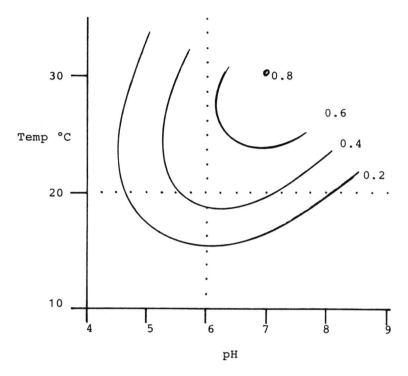

Figure 2 The simultaneous effect of pH and temperature on the analytical solution absorbance.

and a priori knowledge of their effects. In a 2^n experiment where only two levels of each factor are examined, the levels must be sufficiently different to demonstrate clearly the effect of changes in the variable. At the same time, the differences should not be so extreme as to be unrepresentative of the normal acceptable range for the variables.

Returning to our simple three-factor experiment referred to above, the design is traditionally represented in the following manner. In any treatment or factor-level combination, the presence of a variable at its higher, second, level is denoted by writing the factor's lowercase letter in the trial diagram. The absence of any letter in the code indicates that a factor is present at the lower, first, level and the code (1) is used to represent the singular case when all factors are present at their first levels. Thus a is the code for the experimental trial signifying the presence of factor A at its high level with B and C at their low

EXPERIMENTAL DESIGN AND OPTIMIZATION

levels. The set of the eight treatments for the full 2^3 experiment is therefore given by (1) a, b, ab, c, ac, bc, abc.

The statistical analysis of the resultant experimental data is concerned with calculating the main individual effects due to each variable and identifying significant interactions between variables.

The principal effect of any factor is given by the difference between its average response at the upper, second, level and the lower, first, level.

Designating the results obtained from each treatment by the same letter codes, the estimate of the total effects of factor A would be the difference between the sum of trials involving treatment a and the sum of trials not involving this treatment. Thus, the mean effect in the observations of moving from the upper to the lower level of variable A is given by

$$(a + ab + ac + abc)/4 - [(1) + b + c + bc]/4$$

That is,

Mean effect of $A = [a + ab + ac + abc - bc - b - c - (1)]/4$

and similarly

Mean effect of $B = [b + ab + bc + abc - ac - a - c - (1)]/4$

and

Mean effect of $C = [c + ac + bc + abc - ab - a - b - (1)]/4$

Using the same notation, we can define the interaction between two factors as the difference between the observed effect with one factor present along with the upper and lower levels of the second. The AB interaction effect is given by the difference between the effect of A at the higher level of B and the effect of A at the lower level of B. B is tested at its upper level by treatments denoted by b, ab, bc, abc, and the mean effect of A at this upper level is given by

$$(ab + abc)/2 - (b + bc)/2$$

At its lower level of treatment, B is tested by (1), a, c, ac, and the effect of A here is

$$(a + ac)/2 - [(1) + c]/2$$

The mean interaction effect denoted by AB is therefore provided by

$AB = [abc + ab + c + (1) - a - ac - b - bc]/4$

and similarly

$AC = [abc + ac + b + (1) - a - ab - bc - c]/4$
$BC = [abc + bc + a + (1) - b - ab - ac - c]/4$

Table 4 Treatment/effect

Treatment/Effect	Total	A	B	C	AB	AC	BC	ABC
(1)	+	−	−	−	+	+	+	−
a	+	+	−	−	−	−	+	+
b	+	−	+	−	−	+	−	+
ab	+	+	+	−	+	−	−	−
c	+	−	−	+	+	−	−	+
ac	+	+	−	+	−	+	−	−
bc	+	−	+	+	−	−	+	−
abc	+	+	+	+	+	+	+	+
Divisor	8	4	4	4	4	4	4	4

Finally, the mean ABC interaction effect can be derived in a similar manner and is calculated from

$$ABC = [abc + a + b + c - ab - ac - (1)]/4$$

These estimates of the principal factor effects and their interaction are sometimes referred to as *contrasts*. They are linear combinations of the original eight observations and can be tabulated as shown in Table 4.

Each principal effect as derived above is estimated by adding all observations with a plus sign in the appropriate column and subtracting all observations with a minus sign in that column. Note that in such a 2^n factor table each column contains 2^{n-1} plus signs and 2^{n-1} minus signs. In addition to the mean single-factor effects, any interaction column can be obtained by multiplying the signs for the appropriate single-factor columns. In this way the equations above can be obtained directly.

A simple example will serve to illustrate the use of a 2^n experimental design and the analysis of the resulting data. The effect of three instrumental factors on the atomic absorption signal due to copper was examined. The factors, treatments, and observed results expressed in absorbance units, are tabulated below.

A, burner height (mm)	25				10			
B, lamp current (mA)	10		2		10		2	
C, slit width (mm)	0.10	0.05	0.10	0.05	0.10	0.05	0.10	0.05
Absorbance	0.08	0.16	0.12	0.21	0.08	0.16	0.10	0.18
Treatment	abc	ab	ac	a	bc	b	c	(1)

The main effects and interactions can be calculated using the scheme discussed above.

$$\text{Total effect of } \begin{array}{l} A = 0.05 \\ B = -0.13 \\ C = -0.33 \end{array}$$

$$\text{Total interaction, } \begin{array}{l} AB = -0.05 \\ AC = -0.01 \\ BC = 0.01 \end{array}$$

and

$$ABC = 0.01$$

As there is little evidence of interactions, and in view of the small number of tests, the interaction variances can be summed to provide an estimate of the residual experimental (random) error against which we can test the main effects as variance ratios.

Each sum of squares can be calculated by squaring each effect total and dividing by 2^n; in this case, $2^n = 8$.

Source of variation	Sum of squares	Degrees of freedom	F Ratio
Effect A	3.125×10^{-4}	1	3.57
B	2.110×10^{-3}	1	24.11
C	1.360×10^{-2}	1	155.43
Interaction			
AB	3.125×10^{-4}	1	
BC	1.250×10^{-5}	1	
AC	1.250×10^{-5}	1	
ABC	1.250×10^{-5}	1	

From tables, $F_{(0.05,1,4)} = 4.54$.

The results provide evidence that decreasing the slit width (C) and reducing the hollow cathode lamp current (B) significantly enhance the copper absorbance values. Observation height in the burner (A), within the range examined, has little effect on the observed absorbance data.

As the number of considered factors increases, then, the number of treatments to be examined rises rapidly. Consideration of six factors, even if only at two levels, leads to 64 tests, and if replication is employed then a further 64 trials are necessary. It is common for only a single replication to be used in factorial experiments, and the calculation of the factor effects and interactions

Table 5

Trial	Response	I	II	Total	Effect sum of squares (total²/8)	F Ratio
(1)	0.32	0.47	1.29	2.90	1.051	
a	0.15	0.82	1.61	−0.72	0.065	1300
b	0.51	0.73	−0.37	0.50	0.031	620
ab	0.31	0.88	−0.35	−0.04	0.0002	4
c	0.45	−0.17	0.35	0.32	0.013	260
ac	0.28	−0.20	0.15	0.02	0.00005	1
bc	0.53	−0.17	−0.03	−0.02	0.005	100
abc	0.35	−0.18	−0.01	0.02	0.00005	

is achieved using a simplified procedure described by Yates, details of which are provided by Box et al. [3] and Chatfield [4]. The algorithm is applied to the observations after they are arranged and tabulated in the so-called standard order.

Consider the following three-factor exercise. The recorded absorbance of an analytical solution of copper, having reacted with a suitable complexing reagent, is observed to depend on the pH of the solution, its temperature, and the time the solution is allowed to stand before a measurement is recorded. Using solution pH, temperature, and reaction time as three experimental variables examined at two levels, the following results were obtained.

A, solution pH	2				6			
B, temperature (°C)	18		32		18		32	
C, time (s)	20	120	20	120	20	120	20	120
Absorbance	0.32	0.45	0.51	0.53	0.15	0.28	0.31	0.35
Treatment	(1)	c	b	bc	a	ac	ab	abc

Arranging the data in the "standard" order and using Yates's algorithm provides Table 5.

With n factors, n columns should be calculated. Each is produced from the preceding column in a similar manner. The first 2^{n-1} numbers in a column are calculated by adding successive pairs of numbers from the preceding column. The lower 2^{n-1} numbers are then determined from the difference between the same number pairs. Thus in column I of Table 5, the first number is (0.32 + 0.15), the second (0.51 + 0.31), and similarly to the fourth entry (0.53 + 0.35). The next four numbers in this column are then produced from (0.15 −

0.32), (0.31 − 0.51), through to (0.35 − 0.53). Column II is produced from column I by the same process, and finally, the total effects column from column II. The sum of squares for each effect is obtained by squaring the effects total and dividing by 2^n.

To examine each effect and determine which ones have a significant influence on the total variance, it is necessary to have some measure of the random error. As there are expected to be negligible three-factor interactive effects, the sum of squares for this combination can be used to provide an estimate of the residual, experimental error. Using this value, the F ratios are calculated.

From tables, $F_{(0.1,1,1)} = 40$, and it can thus be concluded from the above analysis of the colorimetric data that all three factors examined are highly significant and there is strong evidence of a temperature–time interaction (bc) effect.

Fractional Factorials

A full factorial experimental design provides a complete description of the experiment and all the effects and interactions from the variables studied. As discussed above, however, the full factorial study, while efficient, may prove too expensive or be impracticable, and a 2^n design was examined as providing an estimate of the single-factor effect and the major interactions. Even this much simplified design can be costly in some cases. With n factors, each studied at two levels, the total number of treatments doubles for each factor studied, and if replication is necessary to determine experimental error then the total number of treatments doubles again. Without replication, a seven-factor experiment requires 128 treatments, and eight factors, 256 treatments. One way of reducing the number of trials and treatments is to perform a fractional factorial experimental design. We may know a priori and from practical considerations that certain interactions between factors are impossible, extremely unlikely, or negligible compared with large main effects. Furthermore, if the experiment is a preliminary study, then a high measure of precision in estimating multiple factor interactions may not be required.

Consider the case of a six-factor experiment. A 2^n design dictates that 64 experimental treatments are required to study all main effects and interactions. Of these 64 treatments there are 6 main (single-factor) effects, 15 first-order (two-factor) interactions, 20 second-order (three-factor) interactions, 15 third-order (four-factor) interactions, 4 fourth-order (five-factor) interactions, and 1 fifth-order (six-factor) interactions. If, as is usually the case, the second-order and higher interactions are unlikely to be important, then the maximum information is contained in 21 of the treatments (the main and first-order treatments) from the 64 trials. Therefore 64 treatments have been examined to provide 21 pieces of information, which is wasteful, and a half-factorial experiment could be undertaken, of 32 treatments, which would provide the necessary useful information and be less costly. There are a number of ways of selecting this reduced set of treatments to be investigated, and the method chosen depends on the requirements and purpose of the experiment.

A simple example will illustrate the scheme underlying a typical fractional factorial experiment. Suppose we wish to study a three-factor experiment but use only four treatments. The 2^n factorial design would require the treatments (1), a, b, ab, c, ac, bc, abc and eight trials. To reduce the number of trials, the second-order interaction, abc, can be confounded between two blocks by dividing the trials into those that have an even number of high-level factors and those that have an odd number of factors at the low level, that is,

Block 1: (1) ab ac bc (the even block)
Block 2: a b c abc (the odd block)

Using the four treatments from either block provides four observations, enabling three independent comparisons to be made, and either block will estimate the three main, single-factor, effects,

From Block 1,

Main effect A = $\{(ab + ac) - [(1) + bc]\}/2$
Main effect B = $\{(ab + ac) - [(1) + ac]\}/2$
Main effect C = $\{(ac + bc) - [(1) + ab]\}/2$

In the case of using block 2, then the main effects are obtained directly and the first-order (two-factor) effects are *aliased* in the second-order treatment abc. Assuming the higher-order effects to be negligible, this interaction can be considered to arise from first-order effects.

Further examples of the design and use of fractional factorial experiments to minimize the number of treatments that have to be studied can be found in numerous statistics texts, including reference 2.

6.4 RESPONSE SURFACES AND OPTIMIZATION

For experiments aimed at determining and measuring the effects of changing variables, the factorial methods discussed above can provide a complete picture of the multivariate response. In many cases, however, this complete and overall account of how a system responds to changes in the controlled factors is not always required, or it may be too expensive, in time or materials, to acquire. What is often wanted from a multivariate experimental design is the determination of optimum conditions. What levels of the input factors will give rise to a response simultaneously satisfying some maximum, or minimum, output? It is these questions that are examined and, we hope, answered by the study of response surfaces and optimization techniques.

If the observed response y of an experiment is a function of several controlled continuous variables, $x_1, \ldots, x_k$, then the physical representation of a response surface, the way we may view it, is dictated by the number of variables. For a single variable x, the relationship describing y and x can be represented by a curve. With two variables, $y = f(x_A, x_B)$ is commonly displayed as a contour diagram, each contour line representing a discrete value of the response variable y. With increasing numbers of variables, the display becomes more complicated, although most mathematical search and optimization procedures can operate in the higher dimensional response space defined by these multivariate experiments. Before considering multivariate search techniques, it is instructive to examine the single-factor experiment in more detail.

Single-Factor Optimization

If the observed response variable y is a function of a single controlled variable x, then several methods are available for determining the maximum response. The complete factorial approach would be to make a series of measurements of y for a wide range of levels of x, plot the results on a scatter diagram, and draw a smooth curve through the points. An estimate of the maximum can be visually identified. If $y = f(x)$ is a known or derived, that is, fitted, mathematical function, then an estimate of the maximum response can be obtained by simple calculus. Thus, if a quadratic equation is fitted to the experimental data, using, for example, least squares methods, we have

$$y = a_0 + a_1 x + a_2 x^2$$

and at a turning point the first derivative, dy/dx, is zero. Therefore,

$$a_1 + 2a_2 x = 0$$

and

$$x_{(y=\max)} = -a_1/2a_2$$

This approach usually requires an extensive number of measurements to be made to achieve an accurately modeled representation, and more efficient methods are available. Consider the relationship $y = f(x)$ illustrated in Figure 3. Although the function has not been identified by the experimenter, the practical limits x_L and x_U are known, and it is desired to determine the maximum value of y between x_L and x_U. The most effective search method within a restricted range involves the use of the Fibonacci series of numbers. The first

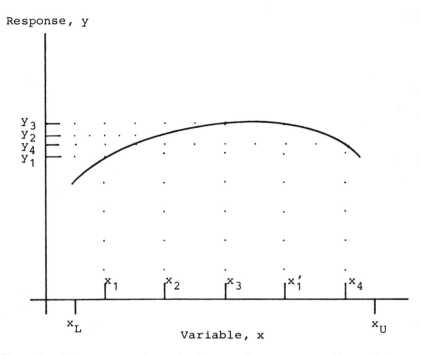

Figure 3 The univariate relationship between the response variable y and the single factor x. X_L and x_U are the lower and upper limits, respectively, of x.

requirement is to decide on the final range of the region that will be accepted as optimal compared with the original search region. Reference to the Fibonacci series indicates the number of experiments required to converge on the optimum y interval. The method is iterative and works by eliminating parts of the search region at each step; therefore, each cycle narrows the region in which the optimum can be present. A detailed discussion of the technique is given in Massart et al. [5]. Unfortunately, the method works effectively only in the absence of experimental error, as excessive measurement noise, random errors, can lead to the exclusion of the required region during the search.

The uniplex method proposed by King and Deming [6] is an alternative but similarly open-ended single-factor search procedure. It is a univariate application of the simplex procedure discussed below. An initial interval between two measured points is made to move by reflections, contractions, and expansions to the optimum point. As no section of the search region is excluded from the experiment, the method can function in the presence of experimental error. With reference to Figure 3, two initial points are selected, $x(1)$ and $x(2)$, and the response $y(i)$, at each is measured. The selection of future test points then follows according to a set of simple rules.

If, as shown in Figure 3, $y(2) > y(1)$, the interval is reflected about $x(2)$ such that the new measurement point is given by

$$x(3) = x(2) + [x(2) - x(1)]$$

with the result in this case that

$$y(3) > y(2)$$

and movement in this direction is accelerated,

$$x(4) = x(2) + C[x(2) - x(1)]$$

With $C = 2$, a typical factor, the result is

$$y(4) < y(3)$$

and we have moved too far in this direction.

A new interval is calculated,

$$x'(1) = [x(2) - x(3)]$$

and the cycle is repeated, starting with the conditions specified at $x(4)$.

As the experiment proceeds, the interval step will contract, and the optimum value of x will eventually be found. When the interval size reduces to some predetermined small value, the procedure is stopped. The complete algorithm for conducting a uniplex search is described in the literature [5,6].

Multifactor Optimization

With more than one controlled variable there is even more incentive to use sequential search designs rather than the complete factorial approach. If only two factors with 10 permitted levels of each are to be examined, 100 measurements must be made to produce a full grid of points to map and construct the experimental response surface. To overcome the demands made by the factorial design, a number of sequential experimental design techniques have been proposed in the literature. The most widely used are those methods based on the measure of steepest gradient and the simplex procedure.

To illustrate the use of these methods in determining optimum conditions to produce a maximum response, consider the hypothetical case illustrated by the contour diagram in Figure 4. The response variable y—for example, solution absorbance—is a function here of two controlled variables x_A and x_B—for example, time and pH. The nature of this response surface is not known to the experimenter, and a suitable starting point is selected as $x_A(1) = 2$ and $x_B(1) = 1$, which provides the response $y(1) = 0.2$.

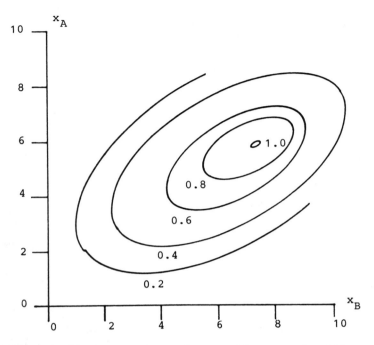

Figure 4 The response of y as a function of the measured variables x_A and x_B

The Method of Steepest Gradient

Provided that the initial measurement is some distance away from the maximum response, as in the case examined here, the response surface about this point can be approximated by a response plane given by the linear equation

$$y = a_0 + a_1 x_A + a_2 x_B$$

The coefficients a_0, a_1, and a_2 can be estimated from a 2×2 experiment centered on $y(1)$. For the example illustrated, the levels of x_A and x_B were examined, and the observed results are as follows. (The results are illustrated in Figure 5.)

Run (i)	Variables		Response
	x_A	x_B	$y(i)$
1	2.0	1.0	0.20
2	1.5	1.5	0.13
3	2.5	1.5	0.30
4	1.5	0.5	0.08
5	2.5	0.5	0.15

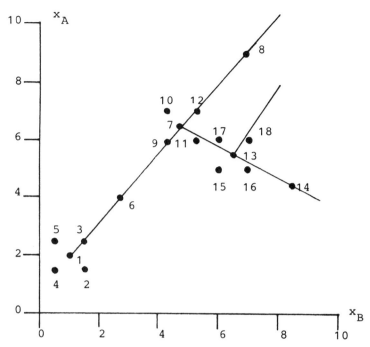

Figure 5 The experimental values (data points) examined and the search directions computed (curve) using the method of steepest gradient for the data shown in Figure 4.

The coefficient a_1 is the change in response when x_A is changed by one unit. For the factorial design shown above, this is half the linear main effect observed between the difference between the averages for the upper and lower x_A levels.

$$a_1 = \frac{(0.3 + 0.15) - (0.13 + 0.08)}{2} = 0.12$$

Similarly for a_2 and the change in response due to x_B:

$$a_2 = \frac{(0.3 + 0.13) - (0.15 + 0.08)}{2} = 0.10$$

and the least squares estimate of a_0 is the average of all five observations in the plane:

$$a_0 = \frac{(0.2 + 0.13 + 0.08 + 0.15)}{5} = 0.172$$

The fitted equation describing the response plane is therefore

$$y = 0.172 + 0.12x_A + 0.10x_B$$

The direction corresponding to the maximum gradient is perpendicular to the contour lines and can be calculated from the fitted equation. Starting at the center of the trial region, given by $y(1)$, the path is followed by simultaneously moving a_2 units in the x_B direction for every a_1 units in the x_A direction. The path is shown in Figure 5. Keeping to this line, the following three new trials were made. Of course, a univariate style optimization could be performed to find the maximum response along this line.

Run	x_A	x_B	y
6	4.0	2.7	0.50
7	6.5	4.7	0.60
8	9.0	6.8	0.28

From the three results, the response $y(6)$ is a considerable improvement on previous observations, and a large step to run 7 is taken. This again shows marked improvement. A further step along the line to run 8 produces a dramatic fall in the value of y, indicating that too large a move has been made. This indicates that further experimentation should be made in the neighborhood of run 7.

Starting then at $y(7)$, a new 2×2 factorial experiment is designed:

Run	x_A	x_B	y
9	6.0	4.2	0.60
10	7.0	4.2	0.45
11	6.0	5.2	0.75
12	7.0	5.2	0.58

Again, if we can assume that these data can be approximated by a plane, the second set of coefficients a'_0, a'_1, and a'_2 can be calculated.

$$y' = a'_0 + a'_1 x_A + a'_2 x_B$$

and, proceeding as detailed above,

$$y' = 0.60 - 0.16x_A + 0.09x_B$$

(The reader may wish to confirm the values of these coefficients.)

Moving along the new line of maximum gradient, two new results are obtained:

EXPERIMENTAL DESIGN AND OPTIMIZATION

Run	x_A	x_B	y
13	5.5	6.5	0.92
14	4.5	8.0	0.48

Although run 13 gives more improvement in the response, a further step along the line to run 14 once again produces a decrease. Returning, therefore, to run 13, the cycle can be repeated.

Run	x_A	x_B	y
15	5.0	6.0	0.85
16	5.0	7.0	0.80
17	6.0	6.0	0.85
18	6.0	7.0	1.00

Fitting a new plane to the data,

$$y'' = a_0'' + a_1'' x_A + a_2'' x_B$$

and determining the coefficients gives

$$y'' = 0.88 + 0.10 x_A + 0.05 x_B$$

The new line is illustrated in Figure 5.

It should be noted that as the treatments approach their optimal values, the assumption that the response surface can be approximated by a simple plane becomes less valid. Instead, the first-order plane model of the surface should be replaced by a second-order quadratic curve to model the more rapidly changing response. More trials will be required to perform the least squares fit to this quadratic approximation.

In practice, the steepest gradient procedure as described is little used in this form because of its slow convergence near the optimum. It is, however, used as the basis for more sophisticated techniques, and detailed discussion is provided by Box et al. [3] and in the more recent texts, including BASIC programs for microcomputers, by Carley and Morgan [7] and Bunday [8].

The Simplex Method

A *simplex* is a geometric figure defined by a number of coordinates, the number of points being equal to one more than the number of factors describing the response surface. Thus, a simplex in two dimensions is a triangle, and one in three dimensions, a tetrahedron. The object of the iterative, sequential search method using the simplex is to force the simplex to move to the region of

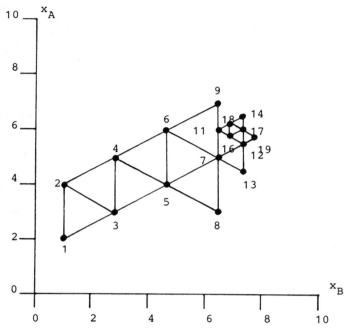

Figure 6 The experimental values (dots) comprising the original simplex and its movement through two-factor space for the data shown in Figure 4.

maximum response, with each new corner of the simplex defining experimental conditions. The algorithm controlling the path of the simplex comprises a set of rules.

The original simplex method was developed by Spendley et al. [9] and modified by Nelder and Mead [10]. The simplex method is elegant and requires calculations that can be performed readily by hand.

Figure 6 illustrates the use of the simplex method with the two-factor example used above. The experiment this time begins with three points defining the vertices of an equilateral triangle, the simplex. It is commonly the case that one edge of the initial simplex is parallel to an axis. From the set of observed responses, the worst value recorded is $y(1)$, and it is logical to expect that the response will increase in the general direction away from this point. To achieve this the simplex is rotated across the edge connecting $y(2)$ and $y(3)$ to produce a reflection at point 4. Points 2, 3, and 4 are considered to define a new simplex, and the search proceeds with the simplex moving quickly across the response surface.

EXPERIMENTAL DESIGN AND OPTIMIZATION

The rules describing the simplex's search, summarized from detailed discussions in Deming [13] and Massart et al. [5] are presented below.

Rule 1. A new simplex, and therefore a move on the response surface, can be made after each additional trial and measurement of the response variable.

Rule 2. A new simplex is produced by rejecting the point of the current simplex giving the worst response and replacing it with its mirror image across the edge defined by the remaining points.

Rule 3. If the newly obtained point gives rise to the worst response of the new simplex, then reflecting back would only serve to produce the preceding simplex. In such cases, the point with the second lowest response is rejected from the preceding simplex, and its mirror image is derived to form the new simplex.

Rule 4. If one point is retained for more than three successive simplexes (i.e., the number of edges of the simplex), the response at this point is determined again. If it remains the highest in the last three simplexes, it is the optimum value obtainable with simplexes of the selected size. A new, smaller simplex must be defined and the cycle repeated.

If on checking the observation at the retained point, the response has been falsely high, then the simplex has become fastened to a false maximum, and this can now be corrected by using the new data and recommencing the cycle.

Rule 5. If a point falls outside a boundary or beyond an axis, an artificially low response is assigned to it and rules 2–4 are applied. This procedure serves to automatically reject an outlying point without stopping the succession of simplexes.

Applying these rules to the example above leads to the trials and observations tabulated in Table 6 (see Figure 6). The first three runs define the initial simplex. The worst response is from run 1, so this is rejected and the new point at $X_A(4)$ and $X_B(4)$ is selected for run 4. This procedure is continued, applying rule 2, giving the observations $y(4)$ through to $y(8)$. The response $y(8)$, however, is worse than $y(5)$ or $y(7)$, and so, by applying rule 3, a new point for run 9 is selected from the reflection of $y(5)$. As the simplex is sticking about $y(7)$, this run is repeated, $y(10)$, and the simplex is reduced to half size to increase its sensitivity to local conditions. This general search procedure is continued, and again a new smaller simplex is soon required as the runs move about the optimum sensitive to the response surface about the maximum; when some predetermined minimum size is reached, the search stops.

Table 6

Run	Simplex	x_A	x_B	y	
1		2.0	1.0	0.20	
2		4.0	1.0	0.25	
3	(1,2,3)	3.0	2.8	0.47	Original simplex
4	(2,3,4)	5.0	2.8	0.52	
5	(3,4,5)	4.0	4.6	0.65	
6	(4,5,6)	6.0	4.6	0.63	
7	(5,6,7)	5.0	6.4	0.85	
8	(5,7,8)	3.0	6.4	0.42	
9	(6,7,9)	7.0	6.4	0.70	Reject second worst point
10	(6,9,10)	5.0	6.4	0.85	Repeat y(7)
11		6.0	6.4	0.83	
12	(10,11,12)	5.5	7.3	0.88	New simplex
13	(10,12,13)	4.5	7.3	0.60	
14	(11,12,14)	6.5	7.3	0.87	
15		5.5	7.3	0.88	Repeat y(12)
16		5.75	6.8	0.96	
17	(15,16,17)	6.0	7.3	0.97	New simplex
18	(16,17,18)	6.25	6.8	0.88	
19	(15,17,19)	5.75	7.7	0.84	

The termination of any iterative, sequential search experiment must be considered and planned in advance of the experiment. For the simpler methods the process can be terminated when either the change in treatments or the change in the observed response is less than some preselected value. In the case of the simplex method, the situation may not be so clear. It is not always apparent whether a true optimum has been reached, particularly if more than two factors are examined. For example, a final simplex may have the same response at each point, in which case the value at the center of the simplex should be considered. Then a decision can be made as to whether this value is to be taken as the optimum or, if it is significantly different from the previous results, whether to reduce the simplex size still further. The chemical literature has abundant examples of the use of these optimization techniques and the terminating criteria found useful in practice. The modified simplex involves more complex rules but still relatively trivial calculations and is more efficient. The so-called supermodified simplex incorporates a quadratic polynomial fit to better estimate treatments but requires more sophisticated computation. The simplex technique is used extensively in analytical science and algorithms, and programs can be found in Carley and Morgan [7] and Zupan [10]. Betteridge and Wade [13] have reviewed the history of simplex optimization.

REFERENCES

1. H. L. Youmans, *Statistics for Chemistry*, C.E. Merril, Ohio (1973).
2. C. J. Brookes, I. G. Betteley, and S. M. Loxston, *Mathematics and Statistics*, Wiley, London. (1966).
3. G. E. P. Box, W. G. Hunter, and J. S. Hunter, *Statistics for Experimenters*, Wiley, New York. (1978).
4. C. Chatfield, *Statistics for Technology*, Chapman and Hall, London. (1978).
5. D. L. Massart, A Dijkstra, and L. Kaufman, *Evaluation and Optimisation of Laboratory Methods and Analytical Procedures*, Elsevier, London. (1978).
6. P. G. King and S. N. Deming, *Anal. Chem.*, 46: 1476 (1974).
7. A. F. Carley and P. H. Morgan, *Computational Methods in the Chemical Sciences*, Ellis Horwood, Chichester. (1989).
8. B. D. Bunday, *BASIC Optimisation Methods*, Edward Arnold, London. (1984).
9. W. Spendley, G. R. Hext and F. R. Himsworth, Technometrics, 4, 441 (1962).
10. J. A. Nelder and R. Mead, Computer Journal, 7, 308 (1965).
11. J. Zupan, *Algorithms for Chemists*, Wiley, New York. (1989).
12. D. Betteridge and A. P. Wade, *Talanta*, 32: 709 (1985).
13. S. N. Deming, Experimental Design: Response Surfaces, in *Chemometrics—Mathematics and Statistics in Chemistry* (B. R. Kowalski, ed.), D. Reidel, New York, p.251. (1984).

FURTHER READING

Anderson, V. L., and R. A. McLean, *Design of Experiments*, Marcel Dekker, New York. (1974).

Miller, J.C., and J. N. Miller, *Statistics for Analytical Chemistry*, Ellis Horwood, Chichester. (1986).

Sharaf, M. A., D. L. Illman, and B. R. Kowalski, *Chemometrics*, Wiley, New York. (1986).

Walsh, G. R., *Methods of Optimisation*, Wiley, London. (1975)

7
Signal Processing and Data Analysis

Barry K. Lavine   *Clarkson University, Potsdam, New York*

Analytical chemistry has become an information science [1]. The microprocessor and the personal computer have made it possible to combine different analytical methods into so-called hyphenated systems. Gas chromatography combined with mass spectrometry (GC-MS) and liquid chromatography combined with mass spectrometry (LC-MS) are examples of sophisticated measurement systems that combine the separation capabilities of chromatography with the capability for compound identification of mass spectrometry. GC-MS and LC-MS have revolutionized the analysis of organic molecules in biological and environmental samples. These techniques also produce large quantities of data. To analyze the larger data sets that are typically generated in studies involving hyphenated systems, analytical chemists have turned to pattern recognition methods. Discriminant analysis, clustering, and principal component analysis are examples of pattern recognition methods that have been used by analytical chemists for data analysis.

Pattern recognition had its origins in the field of image and signal processing. The first study to appear in the chemical literature on pattern recognition was published in 1969 and involved the interpretation of low-resolution mass spectral data [2]. Since then pattern recognition methods have been applied to a wide variety of chemical problems such as chemical fingerprinting [3–6], spectral data interpretation [7–9], and molecular structure–biological activity correlations [10–12]. Over the past two decades, a number of books and review articles on this subject have been published [13–20].

Pattern recognition methods were originally developed to solve the class membership problem. In a typical pattern recognition study, samples are classified according to a specific property by using measurements that are indirectly related to that property. An empirical relationship or classification rule is developed from a set of objects for which the property of interest and the measurements are known. The classification rule is then used to predict the property in samples that were not a part of the original training set.

For pattern recognition analysis, each sample is represented by a data vector $x = (x_1, x_2, x_3, \ldots, x_j, \ldots, x_n)$, where component x_j is the value of the jth descriptor. Such a vector can also be considered as a point in a high-dimensional measurement space. The Euclidean distance between a pair of points in this measurement space is inversely related to the degree of similarity between the objects. Therefore, points representing objects from one class will cluster in a limited region of the measurement space. Pattern recognition is a set of numerical methods for assessing the data structure of this high-dimensional space. The data structure is defined as the overall relation of each object to every other object in the data set.

In this chapter I describe the four main subdivisions of pattern recognition methodology: (1) mapping and display, (2) clustering, (3) discriminant development, and (4) modeling. The procedures that must be implemented in order to apply pattern recognition techniques to chemical problems of interest are also enumerated. Special emphasis is placed on the application of these techniques to problems in profile analysis.

7.1 DATA REPRESENTATION

The first step in a pattern recognition study is to convert the raw data into a computer-compatible form. Normally, the raw data are arranged in the form of a table, a data matrix:

$$\begin{matrix} x_{11} & x_{12} & x_{13} & \cdots & x_{1N} \\ x_{21} & x_{22} & x_{23} & \cdots & x_{2N} \\ \vdots & \vdots & \vdots & & \vdots \\ x_{M1} & x_{M2} & x_{M3} & \cdots & x_{MN} \end{matrix} \qquad (1)$$

The rows of the matrix represent the observations, and the columns are the values of the descriptors. In other words, each row is a data or pattern vector, and the components of the pattern vector are physically measurable quantities called *descriptors*. It is essential that the descriptors encode the same information for all the objects in the data matrix. For example, if variable 5 is the area of a

GC peak for acetaldehyde in object 1, it must also be the area of the GC peak for acetaldehyde in objects 2, 3, . . . ,M. Hence, peak identification is crucial when chromatograms, NMR spectra, or IR spectra are translated into data vectors.

7.2 DATA PREPROCESSING

The next step involves preprocessing. The objective is to enhance the signal-to-noise ratio of the data. In the applications discussed herein, two techniques have been used: normalization and autoscaling. The procedures that should be used for a given data set, however, will be highly dependent upon the nature of the problem.

Normalization involves setting the sum of the components of each pattern vector equal to some arbitrary constant. For GC profile data, this constant usually equals 100, so each peak is expressed as a fraction of the total integrated peak area. In mass spectrometry the peak with the largest intensity is assigned a value of 100, and the intensities of the other peaks are expressed as percentages of this fragment peak. Normalization will compensate for variations in the data due to differences in sample size. However, normalization will also introduce a dependence between the variables that could have an effect on the results of the investigation. Thus, one must take into account both of these factors when deciding whether or not to normalize data [21].

Autoscaling involves standardizing the measurement variables using the z transform so each descriptor or measurement has a mean of zero and a standard deviation of 1, that is,

$$x_{i,\text{new}} = (x_{i,\text{orig}} - m_{i,\text{orig}})/s_{i,\text{orig}} \qquad (2)$$

where $m_{i,\text{orig}}$ is the mean of the original measurement variable and $s_{i,\text{orig}}$ is the standard deviation of the original measurement variable. If autoscaling is not applied, the larger valued descriptors will tend to dominate the analysis. Autoscaling will remove any inadvertent weighting of the variables that would otherwise occur; that is, each variable will have an equal weight in the analysis. Although autoscaling affects the spread of the data (i.e., it places the data points inside a hypercube), it does not affect the relative distribution of the data points in the high-dimensional measurement space.

7.3 MAPPING AND DISPLAY

Graphical methods are often used by physical scientists to study data. If there are only two or three measurements per sample, the data can be displayed as a graph for direct viewing. In other words, the data can be displayed as points in a two- or three-dimensional measurement space. The coordinate axes of the space are defined by the measurement variables. By examining the graph, a scientist can search for similarities and dissimilarities among the samples, find natural clusters, and even gain information about the overall structure of the data set. If there are n measurements per sample ($n>3$), a two- or three-dimensional representation of the measurement space is needed in order to visualize the relative position of the data points in n-space. This representation must accurately reflect the high-dimensional structure of n-space. One such approach is to use a mapping and display technique called *principal component analysis*.

Principal component analysis [22] is a method for transforming the original measurement variables into new, uncorrelated variables called *principal components*. Each principal component (PC) is a linear combination of the original measurement variables. Using this procedure is analogous to finding a set of orthogonal axes that represent the directions of greatest variance (see Figure 1) in the data. (The variance is defined as the degree to which the data points are spread apart in the n-dimensional space.) If a data set has a large number of interrelated variables, then principal component analysis is a powerful method for analyzing the structure of that data and reducing the dimensionality of the pattern vectors.

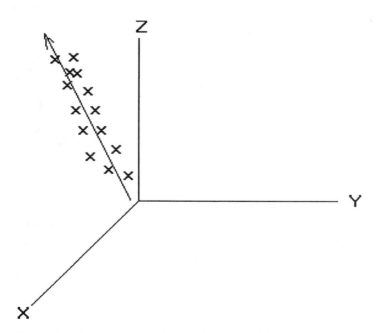

Figure 1 The line through the data points represents the direction of greatest variance in the data. This line defines the first principal component.

The procedure for implementing principal component analysis is as follows. First, the correlation matrix of the data set is computed:

$$C = \frac{1}{N-1} X^T X \tag{3}$$

where X is the autoscaled data matrix, X^T is the transpose of the autoscaled data matrix, and N is the number of columns in X. Next, an eigenanalysis is performed on the correlation matrix. The eigenvector corresponding to the largest eigenvalue represents the direction of greatest variance in the data; each successive eigenvector represents the direction of maximum residual variance. The eigenvectors (i.e., principal components) are then arranged in order of decreasing variance; the first eigenvector is the most informative, and the last eigenvector is the least informative. The two largest eigenvectors or principal components are retained; the values of the first two principal components are computed for each data point. Finally, the points are projected onto a plane defined by the first two principal components. The amount of information in the PC plot relative to the original measurement space is expressed as

$$I = (\lambda_1 + \lambda_2) / \sum_{k=1}^{P} \lambda_k \tag{4}$$

where I is the fraction of the total cumulative variance, λ_k is the eigenvalue of the kth eigenvector, and P is the number of columns in the autoscaled data matrix.

Another approach for solving the problem of representing data points in an n-dimensional measurement space involves using a technique called *nonlinear mapping* [23]. This technique attempts to find a two-dimensional representation of the data that possesses the following attribute: The distances between the data points in the nonlinear map, d'_{ij}, mimic the distances between the corresponding data point pairs in the original measurement space, d_{ij}. In other words, if the data points representing sample j and sample k are close to one another in n-space, they will also be close to one another in the nonlinear map of n-space.

The first step in a nonlinear mapping experiment is to enter estimates for the location of the data points in 2-space. The final outcome of this mapping experiment is highly dependent on the quality of the estimates. It is known that a plot of the first two principal components captures most of the variance or action in the data. Therefore, the starting point for most nonlinear mapping experiments is a PC plot.

The distances between data points in the original measurement space (i.e., n-space), d_{ij}, are then computed using the Euclidean distance metric:

$$d_{ij} = \left[\sum_{k=1}^{v} (x_{ik} - x_{jk})^2 \right]^{1/2} \tag{5}$$

where V is equal to the total number of variables and d_{ij} is the distance between the points representing sample i and sample j. In addition, the Euclidean distance metric is used to compute the distances between the corresponding data point pairs, d_{ij}^*, in the principal component map. Finally, the coordinates of the points in the principal component map are changed so as to minimize the error function (i.e., the so-called mapping error E):

$$E = \sum_{i>j} \frac{(d_{ij} - d_{ij}^*)^2}{d_{ij}^2} \tag{6}$$

The task of minimizing the error function is accomplished by using a gradient method such as steepest descent.

There are other ways of obtaining a visual representation of the data, for example, principal coordinate analysis [24], nonmetric multidimensional scaling [25], and Chernoff's method [26]. However, principal component analysis and nonlinear mapping are the methods most often used by analytical chemists to visualize the structure of multivariate data sets. Principal component plots and nonlinear maps will at times distort the structure of n-space; nevertheless, they should always be generated in order to check the validity of results obtained from a clustering or discriminant analysis experiment. In fact, the combination of principal component analysis or nonlinear mapping with either discriminant analysis or principal component modeling is the best approach to take for tackling a classification problem. In the last section of this chapter, I show how to use mapping and display techniques in tandem with other pattern recognition techniques to solve chemical problems of interest.

7.4 CLUSTERING

Exploratory data analysis techniques are often quite helpful in understanding the complex nature of multivariate relationships. For example, in the preceding section, the importance of using mapping and display techniques for understanding the structure of complex multivariate data sets was emphasized. In this section, some additional techniques will be discussed that also give insight into the structure of a data set. These methods attempt to find clusters of patterns in the measurement space, hence the term *cluster analysis*.

Clustering methods are based on the principle that the distances between pairs of points (i.e., samples) in the measurement space are inversely related to their degree of similarity. Although several different types of clustering algorithms exist (e.g., ISODATA [27], K-Means [28], FCV [29]), by far the most popular is hierarchical clustering [30]. The starting point for this particular algorithm is the distance or similarity matrix. The distances between all pairs of points in the data set are measured, and the resulting similarity matrix is scanned for the smallest value. The two data points corresponding to this entry are combined to form a new point that is located midway between the two original points. (In other words, these two data points are treated as a single cluster of points.) The similarity matrix for the data set is then recomputed. That is, the distances between this new point (i.e., cluster) and every other point in the data set are measured; the rows and columns corresponding to the old data points are removed from the matrix. Again, the similarity matrix is scanned for the smallest value, and the new nearest pair is combined to form a single point. This procedure is repeated until every point has been linked. The result of this procedure is a diagram called a *connection dendrogram* (see Figure 2), which is a visual representation of the relationships between the samples in the data set.

The dendrogram or tree diagram will have information about the number of clusters that are present in the data. However, the interpretation of the data structure on the basis of the dendrogram will very much depend upon the criteria used for assessing similarity. For example, in Figure 2, we can interpret the data

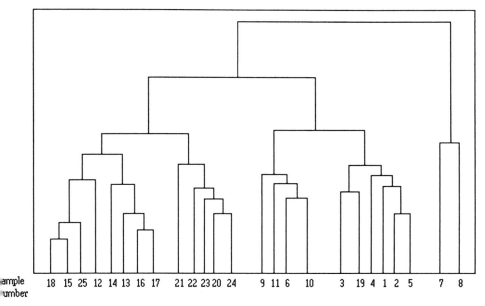

Figure 2 A dendogram obtained from the Marmoset monkey data set [Science, 228: 175 (1985)]. There are four different types of monkeys in the data set: redhead females (sample numbers 1–5), redhead males (6–11), blackhead females (12–19), blackheaded males (20–24).

as being composed of two major point clusters (cluster 1 = samples 18, 15, 25, 12, 14, 13, 16, 17, 21, 22, 23, 20, 24; cluster 2 = samples 9, 11, 6, 10, 3, 19, 4, 1, 2, 5) or four major point clusters (cluster 1 = samples 18, 15, 25, 12, 14, 13, 16, 17; cluster 2 = samples 21, 22, 23, 20, 24; cluster 3 = samples 9, 11, 6, 10; cluster 4 = samples 3, 19, 4, 1, 2, 5). (Samples 7 and 8 are judged to be outliers by the dendrogram.) The criteria for assessing similarity are subjective and will depend to a large degree on the nature of the chemical problem—for example, the goals of the study, the number of clusters that are sought, previous experience, and common sense.

There are a variety of ways to cluster the points using hierarchical techniques (see Figure 3). The single-linkage method [30] assesses the similarity between a data point and a cluster by measuring the distance to the closest point in the cluster. The complete linkage method [30], on the other hand, assesses the similarity between a data point and a cluster by measuring the distance to the farthest point. Average linkage [30] assesses similarity by first computing the

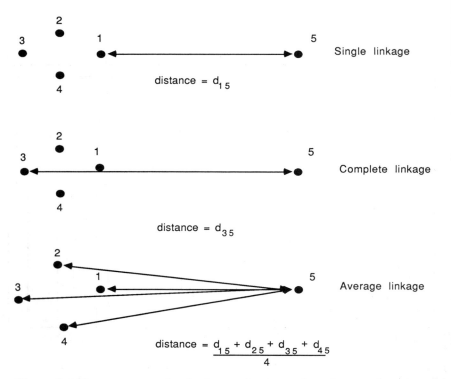

Figure 3 The computation for the distance between a data cluster and a point using (a) single linkage, (b) complete linkage, and (c) average linkage.

distances between all point pairs where a member of each pair belongs to the cluster. The average of these distances is the measure for similarity between a cluster and a data point.

All of these procedures will yield the same results for data sets that have well-separated clusters. However, the results will be different for data sets that have overlapping clusters. Furthermore, clustering algorithms will always partition multidimensional data into clusters, even if the data set in question is composed only of objects from the same class. Therefore, one should analyze a data set using at least two different clustering algorithms. If the clustering results are in agreement, a case can then be made for partitioning the data into distinct subgroups.

7.5 CLASSIFICATION

So far, only unsupervised pattern recognition techniques have been discussed. These techniques analyze data without using information about the class assignment of the samples. Although clustering and mapping and display techniques are powerful methods for analyzing the structure of a data set, they are not sufficient for developing a classification rule. The overall goal of a pattern recognition study, however, is the development of a classification rule that can accurately predict the class membership of an unknown sample. In this section, supervised pattern recognition techniques (i.e., classification methods) will be discussed.

Linear Discriminant Functions

A linear discriminant function can be visualized as a surface dividing a data space into different regions. In the simplest case, that of a binary classifier, the data space will be divided into two regions. Samples that share a common property will be found on one side of the surface, while those samples making up the other category will be found on the other side.

By way of introduction to these binary classifiers, consider Figure 4. The squares represent samples from class 1, and the circles represent samples from class 2. x_1 and x_2 are two experimental variables, for example, the areas of two GC peaks. It is evident from the figure that a line can be drawn through this two-dimensional measurement space such that the squares lie on one side of the line and the circles lie on the other side. Let $d(x) = w_1 x_1 + w_2 x_2 + w_3 = 0$ be the equation of this line, where the w's are the parameters and x_1 and x_2 are the measurement variables (i.e., the areas of the two GC peaks). From the figure it is clear that a sample will belong to class 1 if $d(x) > 0$. On the other hand, a sample will belong to class 2 if $d(x) < 0$. Therefore, $d(x)$ is an example of a linear discriminant function.

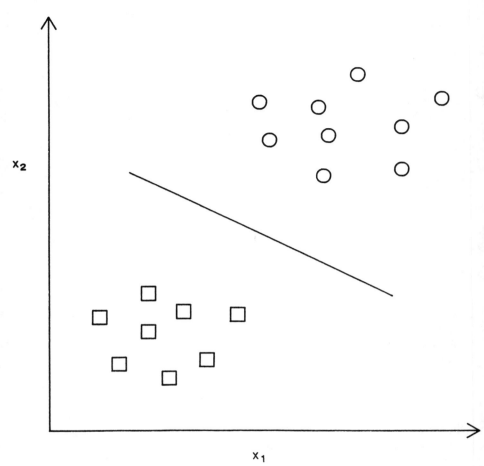

Figure 4 Example of a linear discriminant function application.

A linear discriminant function has the form

$$d(x) = w_1 x_1 + w_2 x_2 + w_3 x_3 + \ldots + w_n x_n + w_{n+1} = W^T x' \quad (7)$$

where

$$W^T = (w_1, w_2, w_3, \ldots, w_{n+1})$$
$$x' = (x_1, x_2, x_3, \ldots, x_n, 1)$$

W^T is called the *weight vector*; it defines the boundary between the two classes. x' is called the *augmented pattern vector*. For discriminant analysis, it is necessary to augment each pattern vector by an extra orthogonal dimension. Because we accomplish this task by appending a 1 to each pattern (i.e., sample) in the training set, the extra orthogonal dimension does not influence the basic properties of the original measurement space; however, the extra dimension does force the decision surface to pass through the origin.

The pattern vector is classified by computing the dot product of the augmented pattern vector and the weight vector [see Eq. (7)]. The distance between the sample and the decision surface is given by the absolute value of the dot product; the sign of the dot product denotes the side of the decision surface on which the sample lies. Therefore, any sample or object in the data set can be classified into its respective class by obtaining the sign of the dot product.

Linear discriminant functions fall into one of two categories: (1) parameteric or probabilistic methods and (2) nonparametric or nonprobabilistic methods. *Parametric methods* are based upon Bayesian statistics and require a knowledge of the data's underlying statistical distribution, that is, the probability density functions. Obtaining these density functions or estimating them for data sets that have less than 500 objects is usually a very difficult task [31]. For this reason, Bayesian methods are seldom applied to classification problems in chemistry. Instead, researchers have turned to nonparametric methods, for example, the linear learning machine [32] and the adaptive least squares algorithm [33], for discriminant development.

Nonparametric methods develop linear discriminants using the class membership and data values of the samples instead of probability density functions or other statistical measures of the data. The linear learning machine or perceptron algorithm is the simplest of these algorithms. It is an iterative method that uses error correction or negative feedback to generate a linear discriminant. The weight vector associated with the linear learning machine is initialized arbitrarily; that is, the elements of the weight vector are assigned arbitrary values. The training set is then examined, one pattern at a time, by the perceptron algorithm. If the pattern vector is classified correctly, the algorithm proceeds to the next pattern and attempts to classify it. On the other hand, if the pattern is misclassified (i.e., $W^T \cdot x' > 0$ but should be less than 0), the weight vector is altered in such a way that it correctly classifies the missed data point. The algorithm then proceeds to the next pattern and attempts to classify it using the altered weight vector. The relationship that is used by the perceptron algorithm to compute the altered weight vector is defined as[1]

$$W'' = W - [2s_i/(x_i' \cdot x_i')] \cdot x_i' \qquad (8)$$

---

[1]The equation for the altered weight vector is expressed as

$$W''' = W + cx_i' \qquad (a)$$

where cx_i' is the multiple of the dot product for the misclassified pattern; that is, cx_i' is a multiple of s_i. The dot product of W'' and x_i' will have the correct sign; that is, the pattern vector will be correctly classified by W'''. *(Footnote continues on p. 222.)*

where W''' is the altered weight vector, W is the weight vector that produced the incorrect classification, x_i' is the pattern vector representing the sample that was incorrectly classified, and s_i is the dot product for the misclassified pattern vector (i.e., $s_i = W \cdot x_i'$). This process will continue until a weight vector capable of classifying all of the patterns in the training set is developed or a preselected number of feedbacks is reached.

If a training set is linearly separable, the linear learning machine will always find a weight vector capable of achieving 100% correct classification. However, the linear learning machine will not find the discriminant, which minimizes the probability of misclassification for a training set that is not linearly separable. In these situations the adaptive least squares algorithm or other gradient descent techniques should be employed for discriminant development.

The adaptive least squares algorithm (ALS) is also an iterative technique that uses error corrective feedback for discriminant development. The weight vector associated with ALS is initialized arbitrarily. After the discriminant scores have been computed, ALS utilizes forcing factors that are based upon the discriminant scores to generate an improved weight vector. The equation for the improved weight vector is

$$W^{(1)} = (X'^T X')^{-1} X'^T L \tag{9}$$

where X' ($n \times p + 1$) is the augmented data matrix, L ($n \times 1$) are the forcing factors (one per pattern), and $W^{(1)}$ is the least squares estimate of the weight vector. (The superscript $^{(1)}$ denotes the first iteration of the algorithm.) If a sample is classified correctly, the forcing factor is equal to the discriminant score. When a sample is misclassified, the forcing factor is equal to the discriminant score plus a correction factor. In other words, for pattern i, $L_i = S_i$ if the sample is correctly classified. L_i is the forcing factor for pattern i, and S_i is the

---

$$W''' \cdot x_i' = s_i'' \tag{b}$$

Combining Eqs. (a) and (b) gives

$$W''' \cdot x_i' = (W + cx_i') \cdot x_i' = W \cdot x_i' + cx_i' \cdot x_i' \tag{c}$$

Substituting s_i for $W \cdot x_i'$ and s_i'' for $W''' \cdot x_i'$ simplifies the above expression to

$$s_i'' = s_i + cx_i' \cdot x_i' \tag{d}$$

Equation (d) can then be rearranged to yield an expression for c:

$$c = (s_i'' - s_i)/(x_i' \cdot x_i') \tag{e}$$

usually, s_i'' is set equal to $-s_i$. Therefore,

$$c = -2s_i/(x_i' \cdot x_i') \tag{f}$$

discriminant score of pattern i. If the pattern is incorrectly classified, $L_i = (S_i + C_i)$. C_i is the correction factor, and

$$C_i = 0.1/(a + d_i)^2 + \text{ß}(a + d_i) \tag{10}$$

where a and ß are constants that are empirically determined and d_i is the distance between the pattern vector and the classification surface (i.e., the discriminant score).

The improved weight vector $W^{(1)}$ is used to compute discriminant scores; forcing factors are again developed from the discriminant scores. Again, a least squares estimate of the weight vector is obtained:

$$W^{(2)} = (X'^T X')^{-1} X'^T L^{(2)} \tag{9a}$$

This procedure is repeated until favorable classification results are obtained or a preselected number of feedbacks has been achieved.

Limitations of Nonparametric Linear Discriminant Functions

Nonparametric linear discriminant functions have provided insight into relationships that are present within sets of chemical measurements. However, these algorithms also have undesirable features. For example, linear discriminant functions are not able to handle the asymmetric case [34–36]. Consider the data space shown in Figure 5. The 1's represent the IR spectra of carbonyl compounds, and the 2's represent the IR spectra of noncarbonyl compounds. It is reasonable to assume that the carbonyl compounds will cluster in a limited region of this space and the other compounds will be distributed randomly throughout the space. This type of data structure, called the *asymmetric case*, is sometimes encountered in binary classification problems and occurs when objects with a well-defined property (e.g., compounds possessing a carbonyl group) are compared to objects that lack this property (e.g., compounds that lack a carbonyl group). In this particular case, a linear discriminant would give arbitrary and incorrect results. In fact, linear discriminants are inherently unreliable for this type of application.

Chance classification is also a serious problem, especially in situations involving nonseparable training sets [37,38]. Using Monte Carlo simulation techniques, it has been shown that the degree of separation in the data due to chance is a function of the number of descriptors, the number of patterns, the class membership distribution, and the covariance structure of the data. Furthermore, these factors do not act independently of one another. This often complicates the interpretation of results obtained in discriminant analysis studies. Therefore, the following procedure has been recommended for assessing the significance of classification scores obtained in real studies.

For a given classification study, 100 data sets consisting entirely of random numbers should first be generated. The statistical properties of the simulated

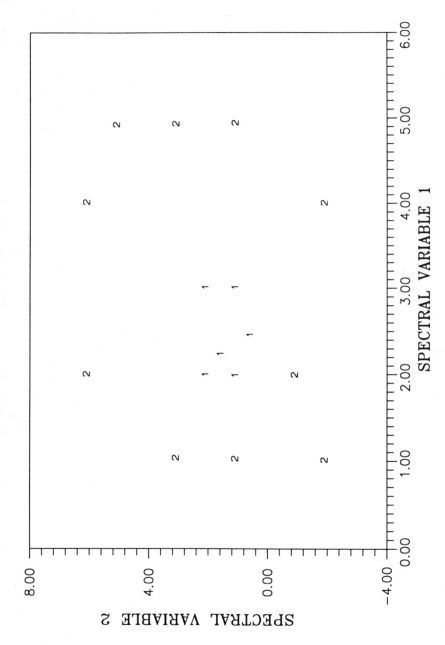

Figure 5 Example of the asymmetric case. The 1's represent carbonyl compounds, and the 2's represent compounds that do not have a carbonyl group.

data (i.e., dimensionality, number of patterns, covariance structure, and the class membership distribution) should be identical to that of the real training set. Next, the separability of each random data set should be assessed. The mean classification success rate for the random data should then be computed and used for comparative purposes. As an example, if the classification success rate obtained in a real study was 80% but the mean score for the simulated data was only 55% and no single trial produced a score higher than 65%, the score obtained in the real study would be judged to be significant. If, on the other hand, the classification success rate obtained in a real study was only 60% but the mean classification success rate for the random data was again 55% and a score of 60% was obtained in at least one trial, the score obtained from this particular study would suggest that there are no discernible differences between the two classes.

SIMCA

Modeling approaches have become more popular in recent years for developing classification rules because of the problems associated with linear discriminant functions. Although there are a number of approaches to modeling clusters, a method based on principal component analysis was developed by Wold and coworkers [39–41] for describing the class structure of a data set. This method is called SIMCA (Soft Independent Modeling of Class Analogy).

In a SIMCA pattern recognition study, a separate principal component analysis is performed for each class in the data set. Principal component models are then developed to approximate the variation in the data. In other words, each class is represented by a PC model that has the form

$$X_i = 1 X_i + T_i \cdot P_i + E_i \qquad (11)$$

where X_i is the $n \times p$ data matrix for class i, $1 X_i$ is the mean vector matrix for class i (i.e., each row in $1 X_i$ is simply the mean vector for class i), T_i is the $n \times F$ score matrix, P_i is the $F \times P$ loading matrix, and E_i is the residual $n \times P$ matrix. The coordinates for each sample in the principal component space are supplied by the score matrix; the loading matrix supplies the necessary information for transforming the original measurement variables into principal components. F is set equal to the number of principal components necessary to model the data. The value of F is either preselected or determined by the procedure of cross-validation (discussed below).

The residual matrix describes how well the model fits the data. From the residuals (the components of the residual matrix), parameters can be derived that describe the data structure. For example, Eq. (12) defines the residual variance of the class, which is a measure of the quality of the category cluster:

$$s_o^2 = \sum_{i=1}^{N} \sum_{j=1}^{P} \frac{(e_{ij})^2}{(P - F)(N - F - 1)} \qquad (12)$$

where P is the dimensionality of the data, F is the number of principal components needed to describe the class, and N is the number of samples in the class. A small value for s_o^2 suggests that these data points cluster in a limited region of the measurement space. How well a sample fits a class model is given by the residual variance of the fit [i.e., Eq. (13)].

$$s_i^2 = \sum_{j=1}^{P} \frac{(e_{ij})^2}{P - F} \tag{13}$$

If s_i^2 is comparable to s_o^2, sample i is representative of the category. On the other hand, if sample i is not representative of the category, s_i^2 will be significantly larger than s_o^2. (An F test is employed to determine whether or not s_i^2 is significantly larger than s_o^2.)

Classification in SIMCA is made on the basis of how well a sample fits the class model. The first step is to compute the coordinates of the sample in the subspace defined by the PC model. Because the loading matrix has already been computed for each PC model, it is a simple matter to compute the coordinates of the sample in the subspace as shown by Eq. (14):

$$t_{ik} = \mathbf{x}_i \cdot P_k^{-1} \tag{14}$$

where t_{ik} is an F-dimensional vector that gives the coordinates of sample i in the subspace defined by the PC model for class k, $\mathbf{x}_i$ is the n-dimensional data vector representing sample i, and P_k^{-1} is the inverse of the loading matrix for class k. A sample is assigned to the class model for which it has the lowest s_i^2 value. [It is a simple matter to compute s_i^2 if t_{ik} and P_k are known. This is accomplished by first subtracting the mean vector of the class and $t_{ik} \cdot P_k$ from $\mathbf{x}_i$. In other words, $\mathbf{x}_i = m_k + t_{ik} \cdot P_k + e_{ik}$, where m_k is the mean vector for class k and e_{ik} is the residual vector. The components of the data vector e_{ik} are then inserted into Eq. (13).] Of course, this assignment is made on the condition that s_i^2 is not significantly larger than s_o^2.

The number of principal components needed for each class model is determined by a procedure called *cross-validation*. The data set is first divided into training set–prediction set pairs. A PC model is then developed for each data cluster in each of the training sets. F is set equal to 1 at the beginning of the run. Samples in the prediction set are then fitted to the appropriate PC model; that is, a sample is fitted to the PC model that represents the class to which it belongs. The residuals are computed and stored away for each sample. The next training set–prediction set pair is then considered for principal component analysis. The process continues until all the training set–prediction set pairs have been analyzed. The entire procedure is then repeated; however, F is now assigned a value of 2. As F approaches the optimum value, the prediction errors (also known as the residuals) decrease. When the optimum value has been exceeded, the prediction errors increase. Thus, the value of F is chosen on the basis of predictive ability, not just "best fit."

SIMCA is a powerful method for analyzing complex multivariate data sets. Although SIMCA requires a greater knowledge of statistics and mathematics than other pattern recognition procedures, it is still a favorite among chemists. Why? SIMCA is able to handle the asymmetric case. It is a simple matter to develop a PC model for the class that forms the compact cluster; unknowns are then fitted to the model. The class assignment for an unknown is then made on the basis of the goodness-of-fit statistic. SIMCA is also able to handle data sets that have a low object/descriptor ratio. During the modeling process, SIMCA extracts F components from the data. Studies [40,42] have shown that SIMCA will function reasonably well as long as F does not exceed one-fourth of the number of samples in the class. Finally, SIMCA is able to characterize the data structure, for example, recognize outliers and assess the compactness of category clusters. Because of these attributes, SIMCA is indeed well suited for developing obscure relationships and for classifying objects.

7.6 APPLICATIONS OF PATTERN RECOGNITION TECHNIQUES

Pattern recognition methods have been applied to a large number of problems in analytical and clinical chemistry. Although the procedures selected for a given problem are highly dependent upon the nature of the problem, it is still possible to develop a general set of guidelines for applying pattern recognition techniques to real data sets. In this section a framework for solving the class membership problem is presented in the context of two recently published studies. In both of these studies biological samples were classified into their respective categories on the basis of their gas chromatographic profiles.

Cystic Fibrosis Heterozygotes versus Normal Subjects

The first study [43–47] involves the application of pyrolysis gas liquid chromatography and pattern recognition methods to the problem of identifying carriers of the cystic fibrosis defect. The biological samples used in this experiment were cultured skin fibroblasts grown from 24 samples obtained from parents of children with CF (4 male, 20 female) and from 24 presumed normal donors (16 male, 8 female). The cells were cultured in modified Eagle's minimum essential medium supplemented with 15% fetal bovine serum and gentamicin. Batches of the growth medium were prepared as needed from a stock solution of modified minimum essential medium. The established cell lines were serially passaged until sufficient material was available for at least four pyrochromatographic experiments.

Pyrolysis of the fibroblast sample was carried out in two stages; first at 400°C and then again at 700°C. Only the pyrochromatograms from the 700°C run were used for the pattern recognition study. A typical CF pyrochromatogram from a 700°C run is shown in Figure 6. The volatile products were separated on a 30-m fused silica capillary column that was temperature-programmed.

For each subject, triplicate pyrochromatograms (PyGCs) were obtained. The 144 PyGCs were standardized (i.e., peak-matched) using an interactive computer program. The PyGCs were divided into 12 intervals using 13 approximately evenly spaced peaks that are always present. The retention times of the peaks within each interval were then scaled linearly for best fit with respect to a reference pyrochromatogram. This peak-matching procedure yielded 214 standardized retention time windows. Therefore, each PyGC was represented as a point in a 214-dimensional space, $x = (x_1, x_2, \ldots, x_j, \ldots, x_{214})$, where x_j is the area of the jth GC peak. The CF chromatographic data set—144 PyGCs of 214 peaks each—was also autoscaled so that each PyGC peak had a mean of zero and a standard deviation of 1 within the entire set of 144 pyrochromatograms.

To apply pattern recognition methods to this data set, the necessary first step was feature selection. For nonparametric linear discriminant functions, the classification results will be significant only if the ratio of samples to measurements is 3 or greater. Therefore, the number of peaks per chromatogram must be reduced to at least one-third the number of independent PyGCs in the data set. For the final results of the analysis to be meaningful this feature selection must be done objectively, that is, without using any class membership information.

For fingerprinting experiments of the type that we are considering here it is inevitable that there will be relationships between sets of conditions used in generating the data and the resulting patterns. One must realize this in advance when approaching the task of analyzing such data. Therefore, the problem is to use the information about the pathological alterations characteristic of CF heterozygotes without being swamped by the large amount of data about the experimental conditions.

In this study it was observed that experimental variables such as cell culture batch number, passage number, donor gender, and column identity can contribute to the overall classification process. For example, a linear discriminant function was developed from the 12 peaks comprising interval 3. The CF PyGCs were linearly separated from the PyGCs of the presumed normal donors. However, when the points from this 12-dimensional space were mapped onto a plane that best represents the pattern space (the plane defined the two largest principal components), clustering on the basis of column identity was observed (see Figure 7). Furthermore, classifiers could be developed from these 12 peaks that yielded favorable results for many of the experimental variables.

Notwithstanding the effects of the experimental variables described above, a discriminant or decision function was developed from the PyGC peaks

SIGNAL PROCESSING AND DATA ANALYSIS

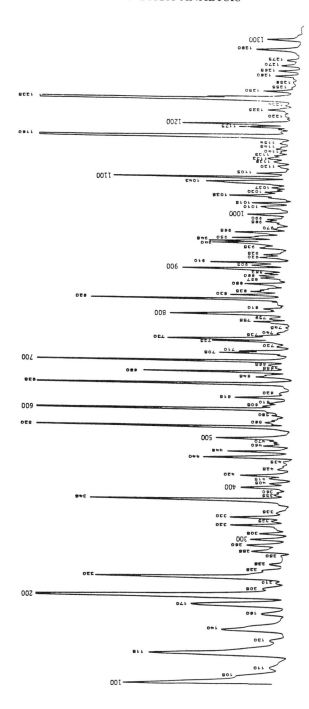

Figure 6 A pyrochromatogram from a CF heterozygote. The major peaks are those with assignments that are multiples of 100. (Courtesy of *Analytical Chemistry*, 57:295–302, 1985)

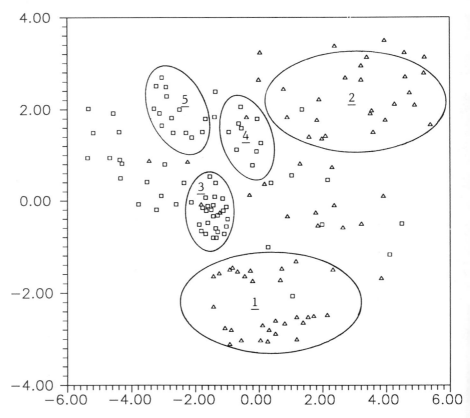

Figure 7 A principal component representation of the measurement space defined by the GC peaks of interval 3. The inverted triangles (normals) and the squares (CF heterozygotes) enclosed by ellipse 1 represent fibroblast samples that were analyzed on the same capillary column. The inverted triangles enclosed by ellipse 2 represent normal samples that were analyzed on another capillary column. (Five different capillary columns were used in the study.) The samples enclosed by ellipses 3, 4, and 5 represent fibroblasts that were grown with different batches of growth medium. (Courtesy *Analytical Chemistry*, 57:295–302, 1985.)

that differentiated between the PyGCs from the CF heterozygotes and presumed normal fibroblasts on the basis of valid chemical differences. The development of this discriminant is described below.

The 65 GC peaks that were present in at least 90% of the PyGCs were used as a starting point for the analysis. The ability of each of these 65 PyGC peaks alone to discriminate between pyrochromatograms from CF heterozygotes and normal cells was assessed. Data-reordering experiments for gender, passage number, and column identity were carried out, and the dichotomization power of each of the 65 PyGC peaks with respect to these experimental variables was evaluated. These reordering experiments showed which PyGC peaks were most subject to experimental conditions, and those peaks were not considered for pattern recognition analysis. In other words, a peak was retained if it had a larger classification success rate for CF/normal than for any other dichotomy. This procedure identifies those peaks that contain the most information about CF versus normal as opposed to the experimental variables. We were attempting to minimize the probability of confounding with unwanted experimental details.

Twelve peaks that produce the best individual classification results when the pyrochromatograms were classified as either CF carrier or normal were used for discriminant development. These 12 peaks spanned the entire pyrochromatogram. A classification rule developed from these 12 GC peaks using SIMCA correctly classified 90% of the PyGCs in the data set. Variance feature selection combined with the linear learning machine and the adaptive least squares algorithm was used to remove peaks that were not relevant to the classification problem. A discriminant that misclassified only eight of the pyrochromatograms (136 out of 144, 94%) was developed using the final set of only six peaks.

The contribution of the experimental parameters to the overall dichotomization power of the decision function developed from the six PyGC peaks was assessed by reordering experiments. The set of PyGCs was first reordered in terms of donor gender, and classification results indistinguishable from random were obtained. Similar studies were performed for passage number and column identity, and comparable results were obtained. The results of the reordering tests suggest that the decision function developed from the six PyGC peaks incorporates mostly chemical information to separate the pyrochromatograms of the CF heterozygotes from normal ones.

The ability of the decision function to classify a simulated unknown sample was tested using a procedure known as *internal validation*. Twelve sets of pyrochromatograms were developed by random selection where the training set contained 44 triplicates and the validation set contained the remaining four triplicates. Any particular triplicate was present in only one prediction set of the 12 generated. Discriminants were developed using the training sets and were tested on the prediction sets. The average correct classification for the prediction set was 87%. This same experiment was repeated except that members of the prediction set included triplicated samples analyzed on the same column or grown in the same batch of growth medium. The average correct classification for the prediction set in this set of runs was 82%. Although the predictive ability of the decision function was diminished when these confounding effects were taken into account, favorable results were still obtained.

Identification of Africanized Honeybees

In the second study [48], gas chromatography and pattern recognition (GC/PR) techniques were used to develop a potential method for differentiating Africanized honeybees from European honeybees. The biological samples used in this experiment were hydrocarbon extracts obtained from 483 adult worker bees. Of the 483 foragers, 67 were heavily Africanized, 63 were moderately Africanized, 145 were designated by the entomologists as simply "Africanized," 178 were European honeybees, and 30 were so-called F1 hybrids. The Africanized honeybees were obtained from colonies in South and Central America. Many of the colonies were designated as moderately Africanized or heavily Africanized by workers at these sites on the basis of the general defensive behavior of the bees. European honeybees were collected from managed colonies maintained in the United States; they represented a variety of commercially available U.S. stocks. The hybrids were obtained from colonies in Mexico where a European queen was allowed to mate freely in a region populated predominantly by Africanized drones.

The following experimental protocol was used to extract the nonisoprenoid hydrocarbons from the bees. First, each bee specimen was soaked in hexane for a period of 72 hr. The hydrocarbon fraction analyzed by capillary column gas chromatography was then isolated from the hexane soaks by means of a silica gel syringe column. Hexane was again used as the eluent. The hydrocarbon fraction was collected and concentrated to dryness under a stream of nitrogen. It was reconstituted with about 50 µl of hexane prior to analysis by capillary column gas chromatography or gas chromatography combined with mass spectrometry (GC-MS).

The hydrocarbon extract was analyzed on a 25-M 5% phenyl methyl silicone fused silica capillary column that was temperature programmed. GC-MS analysis was also performed, and the presence of normal- and branched-chain alkanes and alkenes as well as dienes in the extract was revealed. The GC peaks corresponding to the n-paraffins were then used as retention standards in the capillary column gas chromatographic experiment. Kovat retention indices (KI) were subsequently assigned to the compounds eluting from the column, and these KI values, as well as data from the GC-MS experiment, were used for peak identification.

Each chromatogram contained 65 peaks corresponding to a set of standardized retention time windows. A typical gas chromatographic trace of the nonisoprenoid hydrocarbons from an Africanized forager is shown in Figure 8. The GC experiments were performed on an HP 5890A instrument equipped with a flame ionization detector. The output signal from the flame ionization detector was recorded on an HP 3393A GC integrator.

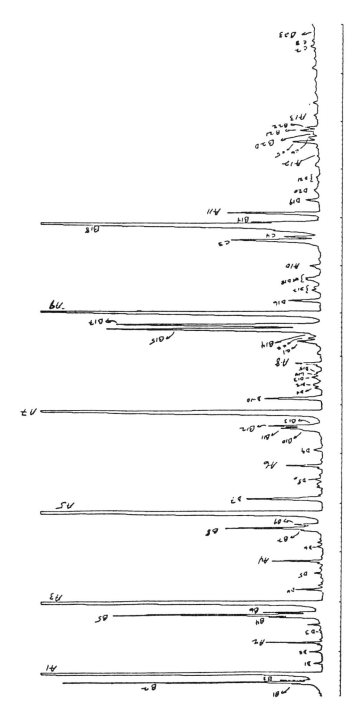

Figure 8 A GC trace of the hydrocarbon extract obtained from the cuticle, wax gland, and exocrine gland of an Africanized forager. A, normal alkanes; B, alkenes; C, dienes; and D, branched-chain alkanes. Peak identification was made on the basis of Kovat retention indices and GC-MS data. (Courtesy of *Microchemical Journal*, 39:308–316, 1989)

Each gas chromatogram was initially represented as a point in a 65-dimensional space, $x = (x_1, x_2, x_3, \ldots, x_j, \ldots, x_{65})$, and where x_j is the area of the jth peak in the chromatogram. However, in this study only the GC peaks corresponding to the unsaturated hydrocarbons (i.e., 31 GC peaks) were considered for pattern recognition analysis. The saturated hydrocarbons were excluded from the analysis because these compounds are believed to be influenced by bee forage. The exclusion of these compounds from the study ensures that information about valid chemical differences between Africanized and European honeybees is isolated from the large amounts of qualitative and quantitative data due to experimental conditions that are also present in the complex chromatograms. This set of data—483 chromatograms of 65 peaks each—was entered into disk storage on a VAX 11/780 superminicomputer. The chromatograms were then normalized to constant sum using the total integrated area of the 31 unsaturated hydrocarbon GC peaks.

Of these 31 unsaturated hydrocarbon GC peaks, only 13 were chosen for pattern recognition analysis. The 13 peaks that were selected possessed a common set of attributes. They were at least moderately well resolved, and computer integration of these peaks always yielded reliable results. The peaks were also readily identifiable in all of the chromatograms, so peak matching was not a problem. Since the feature selection process was carried out on the basis of a priori considerations, and probability of inadvertently exploiting random variation in the data was minimized.

The data were examined using two different types of pattern recognition methodologies: mapping and display methods and discriminant analysis. The pattern recognition analyses were directed toward two specific goals: (1) finding a discriminant that could distinguish both heavily and moderately Africanized bees from European honeybees on the basis of their hydrocarbon profiles and (2) developing the capability of identifying so-called hybrids.

In this study, pattern recognition methods were used to classify the hydrocarbon profiles of the bees according to a particular set of descriptors, the areas of the 13 GC peaks. The test data consisted of 232 chromatograms—130 GC traces of Africanized bees (67 were heavily Africanized, and 63 were moderately Africanized) and 102 GC traces of European honeybees. The data were autoscaled to ensure that each feature (i.e., GC peak) and equal weight in the analysis.

The first step in the study was to use principal component analysis to examine the structure of the training set. In Figure 9, the results of a principal component mapping experiment are shown for the 232 bee specimens. The chromatograms of the bee samples are represented as points in the two-dimensional map. The European honeybees (represented by the 2's in the plot) are well separated from the Africanized bee specimens (represented by the 1's) in the principal component space. The first two principal components account for

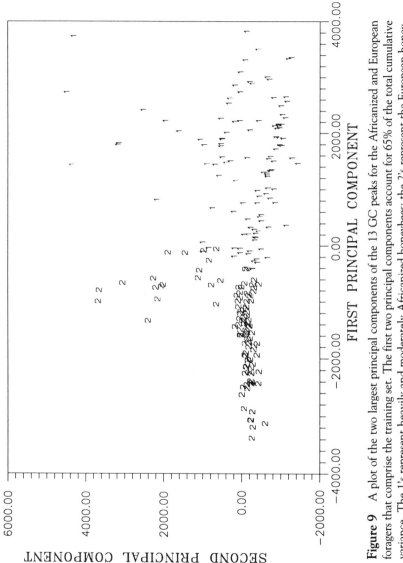

Figure 9 A plot of the two largest principal components of the 13 GC peaks for the Africanized and European foragers that comprise the training set. The first two principal components account for 65% of the total cumulative variance. The 1's represent heavily and moderately Africanized honeybees; the 2's represent the European honeybees. (Courtesy of *Microchemical Journal*, 39:308–316, 1989)

65% of the total cumulative variance. It is important to note that this projection is made without the use of information about the class assignment of the samples. The resulting separation is therefore a strong indication of real differences in the hydrocarbon patterns of these bees as reflected in their gas chromatographic profiles.

Having studied the structure of the data using principal component analysis, the next step was discriminant analysis. A linear discriminant function was developed from the 13 GC peaks for the purpose of separating heavily Africanized honeybees and moderately Africanized honeybees from European honeybees. When the linear learning machine was applied to the 13 GC peaks, it could correctly classify every sample in the training set. The ability of the discriminant to predict the class of an unknown sample was first tested by using a procedure known as internal validation. Eight sets of chromatograms were developed by random selection, where the training set contained 203 chromatograms and the prediction set contained the remaining 29 chromatograms. Any particular bee specimen was present in only one of the eight prediction sets generated. Discriminants were developed using the training sets and tested on the prediction sets. The average classification success rate obtained for the prediction sets was 98%.

To further test the predictive ability of these descriptors and the linear discriminant associated with them, an external prediction set of 251 gas chromatograms was employed. The prediction set contained 145 Africanized foragers, 76 European foragers, and 30 hybrids. Again, a classification success rate of 98% was achieved. All of the European honeybees and 143 out of 145 Africanized honeybee chromatograms were correctly classified. The results obtained for the hybrids were equally encouraging. Twenty-eight of the 30 hybrids were correctly classified; that is, they were assigned to the class composed of samples possessing the African genotype. These results clearly suggest that we can use GC/PR to identify the presence of the African genotype in honeybees.

REFERENCES

1. B. R. Kowalski, *TRACS*, 1: 71 (1988).
2. P. C. Jurs, B. R. Kowalski, and T. L. Isenhour, *Anal. Chem.*, 41: 21 (1969).
3. H. A. Clark and P. C. Jurs, *Anal. Chem.*, 51: 616 (1979).
4. E. Jellum, I. Bjornson, R. Nesbakken, E. Johansson, and S. Wold, *J. Chromatogr.*, 217: 231 (1981).
5. A. M. Harper, in *Pyrolysis and GC in Polymer Analysis* (S. A. Liebman and E. J. Levy, eds.), Marcel Dekker, New York, pp. 373–398 (1985).
6. M. L. McConnell, G. Rhodes, U. Watson, and M. Novotny, *J. Chromatogr.*, 162: 495 (1979).
7. H. Rotter and K. Varmuza, *Anal. Chim. Acta*, 103: 61 (1978).

8. H. B. Woodruff and M. E. Munk, *Anal. Chim. Acta*, 95: 13 (1977).
9. U. Edlund and S. Wold, *J. Magn. Resonance*, 37: 183 (1980).
10. D. R. Henry, P. C. Jurs, and W. A. Denny, *J. Med. Chem.*, 25: 899 (1982).
11. G. K. Menon and A. Cammarata, *J. Pharm. Sci.*, 66: 304 (1977).
12. S. L. Rose and P. C. Jurs, *J. Med. Chem.*, 25: 769 (1982).
13. K. Varmuza, *Pattern Recognition in Chemistry*, Springer-Verlag, Berlin. (1980).
14. B. R. Kowalski, ed., *Chemometrics: Theory and Application*, American Chemical Society, Washington, D.C. (1977).
15. B. R. Kowalski, ed., *Chemometrics, Mathematics, and Statistics in Chemistry*, Reidel, New York. (1984).
16. P. C. Jurs and T. L. Isenhour, *Chemical Applications of Pattern Recognition Techniques*, Wiley-Interscience, New York. (1975).
17. A. J. Stuper, W. E. Brugger, and P. C. Jurs, *Computer Assisted Studies of Chemical Structure and Biological Function*, Wiley-Interscience, New York. (1979).
18. K. Varmuza, in *Computer Applications in Chemistry* (S. Heller and R. Potenzone, Jr., eds.), Elsevier, Amsterdam. (1983).
19. B. R. Kowalski and S. Wold, in *Handbook of Statistics* (P. R. Krishnaliah and L. Kanal, eds.), North-Holland, Amsterdam. (1982).
20. L. Kryger, *Talanta*, 28: 871 (1981).
21. E. Johansson, S. Wold, and K. Sjodin, *Anal. Chem.*, 56: 1685 (1984).
22. I. T. Jolliffe, *Principal Component Analysis*, Springer-Verlag, New York. (1986).
23. J. W. Sammons, Jr., *IEEE Trans. Comput.*, C-20: 68 (1971).
24. J. C. Gower, *Biometrika*, 53: 325 (1966).
25. S. S. Schiffman, M. L. Reynolds, and F. W. Young, *Introduction to Multidimensional Scaling: Theory, Methods, and Application*, Academic, New York. (1981).
26. H. Chernoff, "The Use of Faces to Represent Points in n-Dimensional Space Graphically," Tech. Report No. 71, Department of Statistics, Stanford University, Stanford, Calif. (1971).
27. G. H. Ball and D. J. Hall, "Isodata, a Novel Method of Data Analysis and Pattern Classification," NTIS Report AD699616.
28. J. B. MacQueen, "Some Methods for Classification and Analysis of Multivariate Observations," in Proceedings of 5th Berkeley Symposium on Mathematical Statistics and Probability, (1), Berkeley, Calif.: University of California Press, 1967.
29. J. C. Bezdek, C. Coray, R. Gunderson, and J. Watson, *SIAM J. Appl. Math.*, 40: 358 (1981).
30. D. L. Massart and L. Kaufman, *The Interpretation of Analytical Chemical Data by the Use of Cluster Analysis*, Wiley, New York. (1983).
31. J. H. Han, A. J. I. Ward, and B. K. Lavine, *J. Chem.*, 4: 91 (1990).
32. T. L. Isenhour and P. C. Jurs, *Anal. Chem.*, 43: 20A (1971).
33. I. Moriguchi, K. Komatsu, and Y. Matsushita, *J. Med. Chem.*, 23: 20 (1980).
34. C. Albano, W. Dunn III, U. Edlund, E. Johansson, B. Norden, M. Sjostrom, and S. Wold, *Anal. Chim. Acta*, 103: 429 (1978).
35. W. J. Dunn III and S. Wold, *J. Med. Chem.*, 21: 1001 (1978).
36. W. J. Dunn III and S. Wold, *J. Med. Chem.*, 23: 595 (1980).
37. B. K. Lavine, D. R. Henry, and P. C. Jurs, *J. Chem.*, 2: 1 (1988).

38. B. K. Lavine and D. R. Henry, *J. Chem.*, 2: 85 (1988).
39. S. Wold, *Pattern Recogn.*, 8: 127 (1976).
40. S. Wold and M. Sjostrom, SIMCA—A Method for Analyzing Chemical Data in Terms of Similarity and Analogy, in *Chemometrics: Theory and Practice* (B. Kowalski, ed.), Society Symp. Ser. No. 52, American Chemical Society, Washington, D.C. pp. 243–282 (1977).
41. S. Wold, *Technometrics*, 20: 397 (1978).
42. S. Wold, C. Albano, W. J. Dunn, K. Esbensen, K. Hellberg, E. Johanson, and M. Sjostrom, Pattern Recognition: Finding and Using Regularities in Multivariate Data, in *Food Research and Data Analysis* (H. Martens and H. Russwurm, eds.), Applied Science, London, England. pp. 147–188 (1983).
43. J. A. Pino, "Pyrochromatography of Human Skin Fibroblasts: Normal Subjects vs. Cystic Fibrosis Heterozygotes," Ph.D. Thesis, Cornell University, 1984.
44. J. A. Pino, J. E. McMurry, P. C. Jurs, B. K. Lavine, and A. M. Harper, *Anal. Chem.*, 57: 295 (1985).
45. P. C. Jurs, B. K. Lavine, and T. R. Stouch, *NBS J. Res.*, 90: 543 (1985).
46. B. K. Lavine, "Pattern Recognition Studies of Complex Chromatographic Data," Ph.D. Thesis, Pennsylvania State University, 1986.
47. P. C. Jurs, *Science*, 232: 1219 (1986).
48. B. K. Lavine, A. J. I. Ward, J. H. Han, R. K. Smith, and O. R. Taylor, *Microchem. J.*, 39: 308 (1989).

8
Signal Processing and Data Enhancement

Steven D. Brown *University of Delaware, Newark, Delaware*

In any attempt at collecting data, even with the most advanced instrumentation, the signal that is measured is corrupted with noise. The noise amplitude may be small, and it may not change the signal shape or amplitude significantly, but often the noise contribution can be large, obscuring the true shape and amplitude of the signal.

This chapter is concerned with mathematical methods that are intended to enhance data by decreasing the contribution of the noise relative to the desired signal and by recovering the "true" signal response from one altered by instrumental or other effects that distort the shape of the true response.

Data enhancement can be done in concert with data acquisition, a process that is called *filtering* in this chapter. Many real-time data processing schemes begin with filtering to remove some of the noise, and follow with some sort of decision based on the results of the filtering. The combination of filtering for data enhancement and "online" decisions are useful for process monitoring or instrument control, two areas in which chemometric methods are growing in popularity. Filtering data that are acquired at high rates demand special care because the speed of the data reduction step is very important. The computational burden imposed by a potential method must be considered because the time spent in filtering may decrease the data throughput unacceptably.

Often the data enhancement step is done later, to lower the computational burden placed on the instrumental computer. When data are enhanced "offline," after data acquisition is complete, the process is called *smoothing*. Smoothing methods are much more varied than ones used exclusively for filtering because the time and computational constraints are not so demanding. It should be noted, however, that filtering methods can also be used for smoothing.

Efficient noise removal—by either filtering or smoothing—is a key part of any preprocessing of the data to be subjected to most chemometric methods. There is more to be gained from noise removal than just cosmetic improvement in the data; often, the better the data look to the eye, the more useful the information that can be extracted in subsequent mathematical data analysis. Even though many chemometric methods themselves effect a degree of noise reduction, the presence of significant amounts of noise can cause the failure of mathematical methods that make use of variations in peak amplitude and shape, such as principal component analysis and pattern classification methods. Large amounts of noise can also degrade the results obtained from calibration methods.

8.1 NOISE REMOVAL AND THE PROBLEM OF PRIOR INFORMATION

Consider the noisy data given in Figure 1. Our task is to enhance the signal-to-noise ratio of this spectrum, if we can. To process this spectrum, we must consider what is already known about the data. When a chemical measurement $x(t)$ is obtained, we presume that this measurement consists of signal $s(t)$ corrupted by noise $n(t)$. For simplicity, the additivity of signal and noise is usually assumed, so we describe the measurement by the equation

$$x(t) = s(t) + n(t) \tag{1}$$

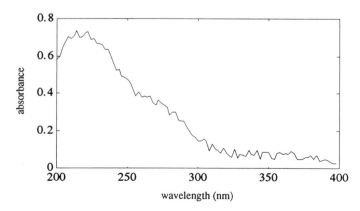

Figure 1 A noisy spectrum to be enhanced.

SIGNAL PROCESSING AND DATA ENHANCEMENT

In this equation, the parameters x, s, and n are all given as functions of time t, but the relation is equally true when x, s, and n are functions of other independent variables, such as potential or wavelength, as in the example here. Collection of data in terms of any of these independent variables can conveniently be taken as done in the time domain, since often (but not always!) these variables are themselves functions of time as a consequence of the instrumental design.

The goal of data enhancement is the extraction of the true signal $s(t)$, given the measured sequence $x(t)$. One measure of the success of this effort is the increase in the signal-to-noise ratio, S/N, where

$$\text{S/N} = \frac{\text{maximum peak height}}{\text{root-mean-square of noise}} \tag{2}$$

Generally this ratio is maximized by attenuating the noise term $n(t)$.

Signal Estimation and Signal Detection

If it happens that $s(t)$, the true form of the signal, is known prior to data enhancement, the process of noise reduction is simplified greatly. In this case, a *known* signal is sought from within the noise. This is a problem in signal *estimation*. The observed quantity x can be taken as a multiple of the known signal plus noise, so that

$$x(t) = as(t) + n(t) \tag{3}$$

The multiplicative factor a is easily found by regression of $x(t)$ onto the signal model given by Eq. (3). Simple application of least squares regression gives

$$a = \frac{\sum_{t=1}^{N} [x(t) - x][s(t) - s]}{\sum_{t=1}^{N} [x(t) - x]^2} \tag{4}$$

The noise is attenuated in the regression. The signal-to-noise ratio can be increased substantially without danger of signal distortion in this case. In fact, once the multiplicative factor a is found, the noise-free version of the observed signal $x(t)$ is easily generated from Eq. (3), taking $n(t)$ as zero.

Problem 1. Given the noisy data sequence

105 18 131 83 140 183 157 80 126 503 857 1405 1651 1476 864 405 313 172 155 131

and the true signal shape

0 0 0 0 0 3 18 77 237 527 852 1000 852 527 237 77 18 0 0

find the signal-to-noise ratio of the noisy data and generate an enhanced data sequence.

If, on the other hand, the signal form $s(t)$ is unknown, it must be identified in the noise. This is an example of signal *detection*. In this case, a suitable model for the desired quantity is not available, and the separation of signal and noise is not as straightforward as with estimation. In attenuating noise, there is a danger of simultaneously altering the characteristics of the signal. Maximizing the S/N ratio may produce significant signal distortion, whereas minimizing signal distortion may improve the observed S/N only a little. A decision must be made as to the relative importance of noise reduction versus signal distortion. Thus, the previous definition of the S/N ratio as a measure of the success of data enhancement must be qualified by the degree to which the true signal $s(t)$ is distorted in the enhancement process.

To examine the distortion of signal and the reduction of the noise, it is necessary to consider the "structure" of the observed data. One especially convenient way of describing the structure of a signal makes use of Fourier series. According to the theory of Fourier series, any signal can be described in terms of a weighted sum of sine and cosine functions of different frequencies [1]. Therefore, finding what frequencies are present in the Fourier series description of an observation provides information on the signal and noise components of the noisy observation.

8.2 FREQUENCY DOMAIN SIGNAL PROCESSING

The Fourier Transform

Any continuous sequence of data $h(t)$ in the time domain can also be described as a continuous sequence in the frequency domain, where the sequence is specified by giving its amplitude as a function of frequency, $H(f)$. For a real sequence $h(t)$ (the case for any physical process), $H(f)$ is series of complex numbers. It is useful to regard $h(t)$ and $H(f)$ as two representations of the *same function*, with $h(t)$ representing the function in the time domain and $H(f)$ representing the function in the frequency domain. These two representations of the same function are called *transform pairs*. The frequency and time domains are related through the Fourier transform equations

$$H(f) = \int_{-\infty}^{+\infty} h(t)\, e^{2\pi i f t}\, dt \qquad h(t) = \int_{-\infty}^{+\infty} H(f)\, e^{-2\pi i f t}\, df$$

The notation employed here for the Fourier transform follows the "system 1" convention of Bracewell [1]. Equations given in this chapter follow this con-

SIGNAL PROCESSING AND DATA ENHANCEMENT 243

vention because it requires fewer 2π factors in some equations. In some texts, Eq. (5) is given in terms of the angular frequency ω, where $\omega = 2\pi f$ (the "system 2" convention of Bracewell), so that any comparisons between equations given here and those given elsewhere should take differences in the notational conventions into account. The units of the transform pair also deserve comment. If $h(t)$ is given in seconds, $H(f)$ has units of seconds^{-1}, or hertz (Hz) (the usual units of frequency), but, as noted above, $h(t)$ may be in any other units. If $h(t)$ is given in units of volts, $H(f)$ will be in units of volts^{-1}.

The Sampling Theorem and Aliasing

The continuous, infinite Fourier transform defined above, unfortunately, is not convenient for signal detection. Most physically significant data are recorded only at evenly spaced intervals in time. In such situations, the continuous sequence $h(t)$ is approximated by the discrete sequence h_n,

$$h_n = h(n\delta), \quad n = \ldots, -2, -1, 0, 1, 2, \ldots \quad (6)$$

where δ is the sampling interval. Associated with the sampling interval δ is a frequency called the *Nyquist frequency*, f_c, which is defined by the relation

$$f_c = 1/2\delta \quad (7)$$

The Nyquist frequency is the maximum frequency that can be present in the continuous sequence $h(t)$ if $h(t)$ is to be perfectly represented by the sampled sequence h_n. In essence, at least two samples must be taken per cycle for a sine wave in order to reproduce that sine wave. A typical reconstruction is shown in Figure 2.

As demonstrated for a simple sine wave, if sampling is done with at least twice the frequency of the highest frequency present in the sampled wave, perfect reconstruction of the continuous sequence from the discrete sequence of samples is possible. This remarkable fact is known as the *sampling theorem*. For any continuous sequence $h(t)$,

$$h(t) = \delta \sum_{n=0}^{N} h_n \frac{\sin[2\pi f_c(t-n\delta)]}{\pi(t-n\delta)} \quad (8)$$

If it happens that the continuous sequence $h(t)$ contains only a finite number of frequencies, that sequence is said to be *bandwidth-limited*. This term is often shortened to the equivalent term *bandlimited*. The sampling theorem states that a signal must be bandwidth-limited to frequencies less than f_c for all signal information to be preserved through the sampling process. If signal information is lost, distortion of the signal will result.

When a sequence that contains frequencies above f_c is sampled at interval δ, a phenomenon called *aliasing* occurs. In aliasing, the sampling process falsely

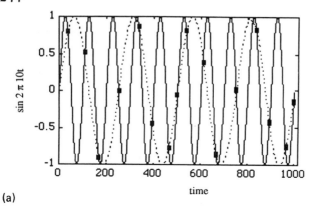

(a)

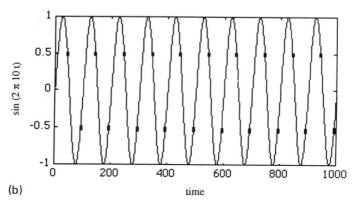

(b)

Figure 2 Reconstruction of a continuous sine wave. (a) Undersampled continuous curve (solid line) with the curve reconstructed from the samples (dotted line). (b) Adequately sampled continuous curve.

assigns all of the power from frequencies outside the frequency range $-f_c < f < f_c$ into that range. To avoid aliasing effects, it is necessary to know the natural (ture) bandwidth of a signal prior to sampling, or to limit the signal bandwidth to a preset value by analog filtering prior to the sampling step. The sampling must then be carried out at such a rate that the highest frequency component of the bandlimited signal is sampled at least twice per cycle. Thus, aliasing limits the number of frequencies that can be examined when the time-domain signal is transformed into the Fourier domain. The effect of aliasing is seen in Figure 2a, where undersampling of the sin $10\pi t$ wave produces an aliased response with apparent frequency $4\pi t$. The power from the $10\pi t$ frequency has been falsely assigned to a lower frequency because sampling was too slow—only frequencies less than $8\pi t$ could be properly represented. When the sampling rate is increased to $20\pi t$, full sampling is possible on the $10\pi t$ frequency, and the sine wave is correctly reproduced, as shown in Figure 2b.

The Bandlimited, Discrete Fourier Transform

The matter of sampling and limited representation of frequencies requires a second look at the representation of data in the time and frequency domains as well as the transformation between those domains. Specifically, we need to consider the Fourier transformation of bandwidth-limited finite sequences of data so that the S/N enhancement and signal distortion of physically significant data can be explored. We begin with an evaluation of the effect of sampling, and the sampling theorem, on the range of frequencies at our disposal for some set of time-domain data.

Suppose a sequence of samples is taken on some continuous function $h(t)$ so that

$$h_k = h(t_k) \quad \text{and} \quad t_k = k\delta \qquad (9)$$

where k is some integer from 0 to $N-1$ and δ is the sampling interval. As we saw above, the Nyquist criterion limits the range of frequencies available from the Fourier transform of h_k. With N samples input, only N frequencies will be available, namely,

$$f_n = n/N\delta \quad \text{with } n = -N/2, \ldots, 0, \ldots, N/2 \qquad (10)$$

By convention, this range of frequencies, where n varies from $-N/2$ to $N/2$, is mapped to an equivalent set running from 0 to N. Because of this mapping, zero frequency now corresponds to $n = 0$, positive frequencies from $0 < f < f_c$ correspond to values $1 \leq n \leq N/2-1$, and negative frequencies $-f_c < f < 0$ correspond to values $N/2 + 1 \leq n \leq N-1$. The alert reader will note that in the Nyquist mapping of frequencies the value $n = N/2$ corresponds to both $-f_c$ and f_c; in fact, sampled sequences are *cyclic*, being joined at the extreme ends.

This is why, no matter what mapping is used, only N frequencies are available from N samples, and $N/2-1$ of these frequencies will be negative frequencies.

The integral transforms given in Eq. (5) can now be approximated by discrete sums, so that the Fourier transform pairs now are described by the equations

$$H_n = \sum_{k=0}^{N-1} h_k \, e^{2\pi i k n / N} \tag{11a}$$

$$h_k = 1/N \sum_{n=0}^{N-1} H_n \, e^{-2\pi i k n / N} \tag{11b}$$

Calculation of these discrete Fourier transforms (DFT) is not at all rapid if done directly. Transformation of a small data set, say 200 samples, might take several minutes to an hour on a laboratory computer if Eqs. (11) were programmed and the transformations attempted. If, on the other hand, the data are sampled so that N is a multiple of 2, a fast Fourier transform (FFT) algorithm can be used to perform the above transform operations in about 1 s, permitting a significant savings in time. For this reason, the FFT algorithm is most often used to accomplish the DFT process. A number of FFT algorithms now exist; two popular ones are the Cooley–Tukey transform and the related Sande–Tukey transform. The mechanistic details of these transforms are covered in many textbooks. Good discussions of FFT algorithms are provided by Bracewell [1] and by Press et al. [2]. Other fast methods for obtaining the Fourier transform, some of which do not require data set sizes to be multiples of 2, are discussed by Bose [3].

Shown in Figure 3 are the plots of H_n for the spectral data given in Figure 1. Note that the units of the abscissa are wavelength^{-1}, the units of "frequency" for this member of the transform pair. Note also the negative frequencies, which give the "mirror image" effect to the transform.

Many FFT algorithms store data so that the frequencies run from zero to the largest positive value, then from the largest negative value to the smallest negative value, as this improves calculation speed. In these, a "mirror image" effect is noticeable when the frequency domain data are plotted, but the plots differ from those shown in Figure 3: Now the left half of the plots shown in Figure 3 is appended to the right half. Recalling the cyclic nature of sample sequences will help the reader in realizing that both ways of describing the frequency sequences are equivalent.

Properties of the Fourier Transform

The interconversion of transform pairs by the Fourier transformation has several unique properties that will be important in analyzing the structure of signals and noise. For the transform pairs $h(t)$ and $H(f)$, important relations are summarized in Table 1. The reader should take note that even though continuous functions

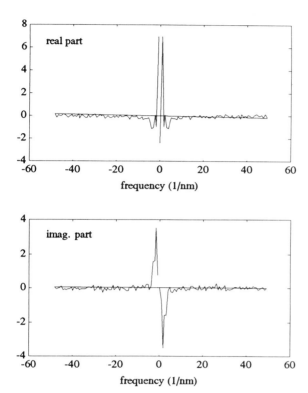

Figure 3 Fourier transform of the noisy spectrum in Figure 1.

are used to illustrate the properties of the transformation, all of the properties listed in the table are identical for the continuous and discrete Fourier transforms.

The first property listed in Table 1, called the *similarity property*, concerns the scaling of the transform pairs. Because the transform pairs describe the same function, the scaling described by the similarity property can refer to "time scaling," as given in Table 1, or it can describe "frequency scaling," namely

$$\frac{1}{|b|} h(t/b) \Leftrightarrow H(bf) \tag{12}$$

Similarly, the shift property can describe "time shifting," as given in Table 1, or it can describe "frequency shifting," where

$$h(t) \, e^{-2\pi i f_0 t} \Leftrightarrow H(f - f_0) \tag{13}$$

Table 1 Some Properties of the Fourier Transform[a]

Property	Time domain {h(t)}	Frequency domain {H(f)}
Similarity	$x(at)$	$1/\|a\| \; X(f/a)$
Translation (shift)	$x(t-a)$	$e^{-2\pi i a f} X(f)$
Derivative	$x'(t)$	$2\pi i f X(f)$
Symmetry	$x(-t)$	$X(-f) = X(N-f)$
Complex conjugate	$x^*(t)$	$X^*(-f)$
Power (Parseval's theorem)	$\sum_{t=0}^{N-1} x(t) \cdot x^*(t)$	$\sum_{t=0}^{N-1} X(f) \cdot X^*(f)$
Linearity	$a\{x(t)\} + b\{y(t)\}$	$a\{X(f)\} + b\{Y(f)\}$
Convolution	$x(t) \cdot y(t)$	$X(f)Y(f)$
Autocorrelation	$x(t) \cdot x^*(t-a)$	$X^*(f)X(f) = \|X(f)\|^2$
Correlation	$x^*(t) \cdot y(t-a)$	$X^*(f)Y(f)$

[a] a and b are arbitrary constants. The symbol $\cdot$ used between variables denotes a convolution operation, so that the notation $a \cdot b = \sum_{k=1}^{N} a(k)b(r-k)$. When used as a superscript, an asterisk denotes complex conjugation, so that $(a + ib)^* = (a - ib)$.

Parseval's theorem demonstrates that the total power is the same whether it is computed in the time domain or the frequency domain. It is often of interest to determine the power in some small interval $f + df$, as a function of the frequency. This is called the *spectrum* of $h(t)$.

For the pairs of functions $\{x(t), y(t)\}$ and $\{X(f), Y(f)\}$, which are related by the Fourier transform, other unique properties exist. These are listed in the lower half of Table 1 and are briefly summarized below.

The *addition property* demonstrates that the sum of functions is preserved in the transformation. This, together with the similarity property, indicates that the Fourier transformation is a linear operation.

One of the most useful relations in signal processing is the *convolution property*. The convolution of two discrete (sampled) functions $x(t)$ and $y(t)$ is defined as

$$x(t) \cdot y(t) = \sum_{k=0}^{N-1} x(k) y(t-k) \tag{14}$$

where $x(t) \cdot y(t)$ is a function in the time domain and $x \cdot y = y \cdot x$. According to the convolution property, the convolution of two functions has a transform pair that is just the product of the individual Fourier transforms. A property related to convolution is cross-correlation. The *cross-correlation* of two functions $x(t)$ and $y(t)$, sampled over N points, is defined as $\gamma_{xy}(\tau)$, where

$$\gamma_{xy}(\tau) = \text{corr}(x,y) = \sum_{k=0}^{N-1} x^*(k) \, y(k+\tau) \tag{15}$$

where τ is called the lag. This function is related to its transform pair by the correlation property. The cross-correlation of x and y has a transform pair that is just the product of the complex conjugate of the individual Fourier transform of one function multiplied by the Fourier transform of the second function. A special case of the correlation property arises when a function is correlated with itself, an operation called *autocorrelation*. In this case corr(x,x) has the transform pair $|X(f)|^2$.

As noted above, the properties of convolution and correlation are the same whether or not a continuous or discrete transformation is used, but, because of the cyclic nature of sampled sequences discussed above, the mechanics of calculating correlation and convolution of functions are somewhat different. The discrete convolution property is applied to a periodic signal s and a finite, but periodic, sequence r. The period of s is N, so that s is completely determined by the N samples $s_0, s_1, \ldots, s_N$. The duration of the finite sequence r is assumed to be the same as the period of the data: N samples. Then, the convolution of s and r is

$$s * r = \sum_{k=-N/2+1}^{N/2} s_{j-k} r_k = S_n R_n \tag{16}$$

where S_n is the discrete Fourier transform of the sequences s, and R_n is the discrete Fourier transform of the sequence r. The sequence r_k maps the input s to the convolved output channel k. For example, r_1 tells the amount of signal in sample j that is placed in output channel $j+1$, while r_3 maps input signals in sample j to output channel $j+3$, and so forth. Figure 4 demonstrates the convolution process.

Note that in Figure 4 the response function for negative times (r_{-1} and r_{-2} here) is wrapped around and is placed at the extreme right end of the convolving sequence. To avoid an effect called *convolution leakage*, it is important to keep in mind the requirements for convolution: Sequences s and r have the same length, and both are treated as periodic. The latter requirement means that the convolution operation will be cyclic, and we must take care to protect the ends of sequence s. To do so, we must ensure that the first and last k elements of sequence s are zero, where k is the *larger* of the number of nonzero negative elements of sequence r or the number of nonzero positive elements of sequence r. This "padding" of the sequence to be convolved avoids convolution leakage (which occurs in the padded areas, but these are of no interest to us and can be discarded). The effectiveness of padding is illustrated with the exercise below. Further discussion of the details of discrete convolution is given in texts by Bracewell [1], Bose [3], and Press et al. [2].

Problem 2. Given the sequence r and the padded sequence s as indicated below,

r: 2 1 0 0 0   0 0 0 0 0 0 0 0 1
s: 0 0 2 0 0 −2 0 0 0 2 2 2 2 0 0

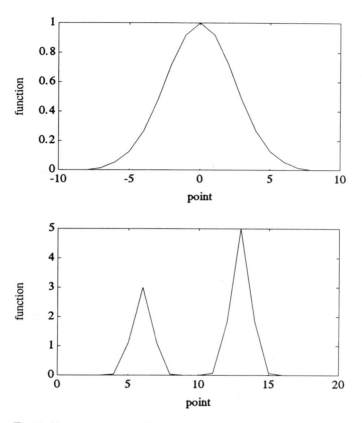

Figure 4 Discrete convolution of two functions. The top function is convolved with the middle function to produce the result shown by the dotted line.

a. Perform the discrete convolution of r and s, using the exact sequences given.
b. Now decrease the size of s by removing the first and last two elements of s. Do the same for r, but take out four zero elements from the center. Again perform the discrete convolution operation. Compare the results of this convolution with those obtained in (a). The difference is due to "leakage" of one end of the cyclic operation into the other end.

With a background in the mechanics and the properties of the Fourier transform, we are now ready to investigate ways of enhancing signal and removing noise while minimizing signal distortion. Two approaches are possible. We could deal with the signal directly, in the time domain, or we could take advantage of the simplicity of some operations in the frequency domain in processing the signal there, after a suitable Fourier transformation. Both methods are commonly used in the processing of chemical data. Each is discussed below.

8.3 FREQUENCY DOMAIN SMOOTHING

Smoothing

Consider again the noisy time domain signal given in Figure 1. For the examples discussed below, the true signal $S(t)$ is unknown. We would like to improve the S/N ratio of this signal. In the time domain, noise and signal seem to coexist, but in the frequency domain the signal exists as a set of low frequencies, and the noise as a set of high frequencies, as shown in Figure 3. To enhance the signal, we need only decrease the amplitude of the noise-containing frequencies while leaving the signal-containing frequencies unchanged.

Figure 5 shows one way to reduce noise without altering the signal significantly. If the frequency-domain representation of the noisy signal is multiplied by a "window" function, the resulting frequency-domain representation has zero amplitude at frequencies where the window function is zero, and unchanged amplitude at frequencies where the window function is 1. Picking the window transition point (where the function changes from 1 to 0) to be in the region between signal frequencies and noise frequencies and multiplying the noisy

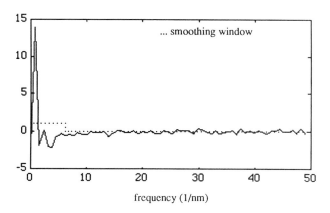

Figure 5 Window function (dotted line) and the real part of the Fourier transform. The window is applied to both real and imaginary parts of the transformed data. The window cutoff is indicated at frequency point 7.

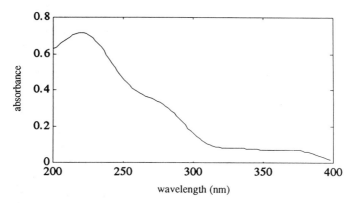

Figure 6 Smoothed data after inverse transformation.

frequency-domain representation by this window function will result in an enhanced frequency-domain representation, which can then be converted back to a time-domain representation, as shown in Figure 6. The enhanced signal has considerably less noise than was contained in the original signal. This signal has been *smoothed* by the application of the window function in the frequency domain. The window function is called the *transfer function* of the smoother.

A very important consideration in the method described above is the location of the *cutoff frequency* for the transfer function. This point is the frequency where the transfer function changes from a value of nearly 1 to a value of nearly zero; frequencies above the cutoff frequency are mainly attenuated, while frequencies below this point are mainly passed through the filtering operation. In the example presented above, it was easy to see where the cutoff frequency should be placed, because signal and noise were well separated in the frequency domain. This convenient separation of signal and noise in the frequency domain is not always true, however. Noise and signal often overlap in the frequency domain as well as in the time domain.

Smoothing with Designer Transfer Functions

When significant overlap of noise and signal occurs in the frequency domain, the successful enhancement of the data depends on the selection of a suitable smoother transfer function. It is generally not satisfactory to zero part of the frequency-domain representation while leaving other parts intact, as was done in the example above. Usually, this smoothing operation will distort the signal. One reason that this method fails can be seen from the properties of the impulse response function used in the time-domain convolution. As shown in Figure 7,

SIGNAL PROCESSING AND DATA ENHANCEMENT

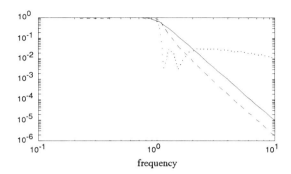

Figure 7 Examples of designer transfer functions. The Butterworth (solid line), Chebyshev (dashed line), and elliptic (dotted line) filters are shown for the fifth order. A log-log plot is used here to enhance differences in the transfer functions.

the transform pair for a step-shaped transfer function produces an impulse function of the form $(\sin x)/x$. When convolved with a noisy signal, this function produces spurious oscillations in the smoothed data. The oscillations, or "feet" on a peak, arise from the abrupt change made in a step-shaped transfer function [4–6].

To avoid apurious oscillations, a designer transfer function can be crafted to obtain a specific type of smoothing. Some common designer transfer functions are based on Chebyshev polynomials, elliptic functions, and the Butterworth filter equation [3,7]. Examples of these transfer functions are shown in Figure 7. These particular transfer functions are conveniently implemented in hard-wired circuits. Other designer transfer functions can be made empirically. For example, a trapezoidal function might be devised to smooth a noisy spectrum, as shown in Figure 8. This smooths data in a way very similar to that done by the step-shaped transfer function, but it reduces the slight, spurious oscillations on the peaks in the smoothed data, as demonstrated in Figure 9. Although the oscillations are quite small in this example, and the use of a trapezoidal window is not essential, often these oscillations can be much larger, particularly when the cutoff point is selected in a region where the frequency-domain representation has nonzero values. Altering the transfer function in this way is sometimes called *apodization* because it removes spurious "feet" from the smoothed data [5,6].

If the true signal shape is known before smoothing, a special designer transfer function can be created. In this case, the transfer function can be the complex conjugate of the frequency response of the signal itself. The result is a *matched smoother*, because the transfer function is matched to the characteristics of the signal. When the complex conjugate of the Fourier transform of the

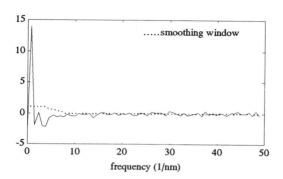

Figure 8 Smoothing in the frequency domain with a trapezoidal function. Here a trapezoid was used with vertices at frequency points 6 and 12.

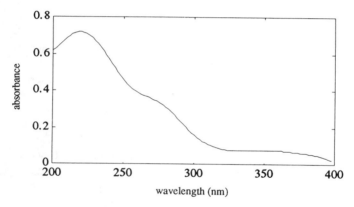

Figure 9 Smoothed data from Figure 8.

desired signal shape is used as the transfer function, the smoother operation is equivalent to a time-domain cross-correlation of the noisy signal with the true, noise-free signal, which, at lag 0, reduces the noise without altering signal shape [8]. This smoother permits recovery of the desired signal from large amounts of noise, just as in the regression example given earlier. In fact, the correlation operation is equivalent to the regression method outlined above. Figure 10 demonstrates the noise removal that is obtained with matched smoothers. Note, however, that a *convolution* of the noise-free and noisy signals would not be useful, because the convolution alters the smoothed signal shape.

Problem 3. Smoothing is generally done after removal of negative frequencies. Strictly speaking, when the negative frequencies are discarded, it is appropriate

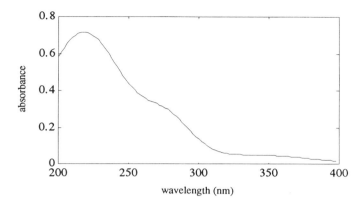

Figure 10 Results from matched smoothing in the frequency domain.

to multiply the positive frequencies by 2. Zero frequency is left unchanged. Why is this correction necessary? What will the resulting time-domain signal obtained upon back-transformation of the truncated frequency-domain data look like if this correction is not done?

Problem 4. The cyclic nature of sampled transforms must always be kept in mind when dealing with Fourier transforms. For example, a direct transformation of the noisy data in Figure 1 does *not* give the result shown in Figure 3. Instead, a preprocessing step, consisting of the subtraction of a line drawn between the two ends of the noisy spectrum, is done prior to taking a transform. In this way, both ends of the corrected spectrum have the same value. Explain, using the convolution property of the transform, why this preprocessing step is necessary.

8.4 TIME-DOMAIN FILTERING AND SMOOTHING

Smoothing

It is also possible to smooth data in the time domain. As shown above, multiplication of two functions in the Fourier domain is exactly equivalent to convolution of their transforms in the time domain. Thus, rather than using a multiplication of two frequency-domain representations, we will need to convolve two time-domain representations. One representation is simple: It is the noisy time-domain signal given in Figure 1. The other, according to the properties of the Fourier transform, should be the time-domain representation of the transfer function shown in Figure 4. This function, which is the smoother *impulse response function*, is shown in Figure 11. The discrete convolution of the

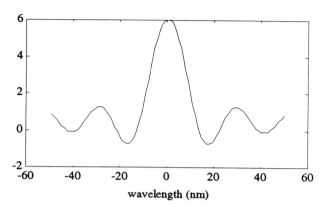

Figure 11 Impulse response function. Note that the Fourier transform of a step function (the smoother window function) is of the form (sin x)/x.

noisy signal with the time-domain impulse response function, defined by the equation

$$s(t) = h(t) * x(t) \tag{17}$$

is shown in Figure 12. Comparison of Figures 9 and 12 confirms the equivalence of the time-domain and frequency-domain approaches.

Because multiplication is simpler to perform than the convolution operation, it is easier to perform this smoothing in the frequency domain. Having a fast transform between the two domains is an important consideration, because smoothing a time-domain signal will require two transformations: one to convert the time-domain signal to the frequency domain, and one to convert the results of the smoothing back to the time domain. Use of the FFT or another fast transform algorithm generally makes this consideration unimportant, except in real-time applications, where very high data throughput is essential.

Filtering

When very rapid enhancement of data is desired, it might be necessary to enhance the data as they are collected. For this reason, the data enhancement step is carried out in the time domain, both to avoid spending the time needed for the two transformation processes and to avoid any delays due to the need for complete observations in any data enhancement method based on smoothing. Real-time enhancement of incomplete data is called *filtering*, and it is always done in the time domain. If the data were analyzed in the frequency domain, complete observations would be needed to prevent problems in representing all signal frequencies; the number of frequencies available cannot exceed the number of time-domain data. In time-domain filtering or smoothing, the basic

SIGNAL PROCESSING AND DATA ENHANCEMENT 257

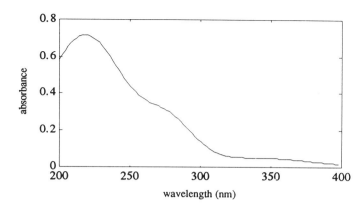

Figure 12 Time-domain smoothing of the noisy data in Figure 1 with the impulse response function of Figure 11.

operation is the cyclic convolution of the filter (or smoother) impulse response function with the noisy data to generate a signal with attenuated noise characteristics. As discussed above, the cyclic convolution operation is more commonly expressed as a discrete sum over the number of measurements available, N, so that the kth filtered datum is determined by the relation

$$s(k) = \sum_{n=1}^{N} h(n)x(n - k) \tag{18}$$

where $h(n)$ is the filter impulse response function defined over the set of N points and the input data are given by the sequence $x(n)$. This difference equation is the defining relation for a *finite impulse response* (FIR) filter, so named because the filter impulse response function has finite (meaning noninfinite) values everywhere. With FIR filters the present response $s(k)$ is not dependent on previous values of s [3,9].

To discover the way in which FIR filters enhance data, consider the simple FIR filter given by the impulse response equation

$$h(n) = 1/N \tag{19}$$

This impulse response defines the simple moving average filter, sometimes known as the *boxcar averager*. As discussed above, the convolution operation defined by Eq. (18) can be regarded as a moving of the impulse response function $h(n)$, which operates over a predetermined "window," through the data $s(n)$. Here the impulse response is a simple average of the values of the noisy signal $x(n)$ inside an N-point "window," as shown in Figure 13.

Averaging reduces the effects of noise, if the noise is fairly random [10], and the moving-average filter removes high-frequency noise well. It is less successful

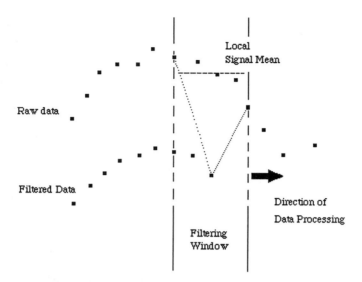

Figure 13 FIR signal processing with a moving-average filter.

at removing low-frequency noise because these variations are less likely to be affected by the averaging. It also is not able to filter the extreme "ends" of the data set, since the window of data is averaged to estimate a new value for the *middle point of the window*. Starting a five-point window in a spectrum means that the first two points cannot be improved; similarly, the last two points are never reached by the moving window. Typically, for a window of $2N+1$ points, $2N$ points will be left unfiltered. These unfiltered points can be seen at the ends in Figure 14, where the operation of the simple five-point moving-average filter is demonstrated on the noisy data of Figure 1. This fast operation, which is easily done in real time, has effectively removed the high-frequency noise, but some of the low-frequency noise remains, as does a slight offset term.

The size of the filter window determines the number of points averaged by the filter. As one might expect, the number of data averaged has a profound effect on the noise reduction process. It also strongly affects the amount of distortion of the enhanced data. The effect of window size on noise reduction and on distortion is best seen through examination of the filter transfer function. Three of these functions, for different window sizes, are shown in Figure 15. Clearly, choice of window size drastically changes the number of frequency components that pass the filter. Larger windows pass a considerably smaller range of frequencies. The collapse of the filter transfer function as the filter window is increased has some disadvantages, too. A large window size may give a filter transfer function that does not pass frequencies associated with signal.

SIGNAL PROCESSING AND DATA ENHANCEMENT

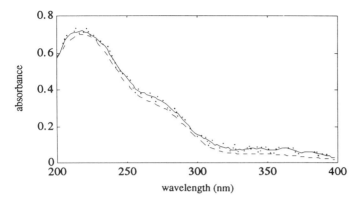

Figure 14 Time-domain processing of noisy data with a 10-point moving-average filter. The noisy data are shown in a dotted line, and the moving-average filtered data as a solid line. The true signal is shown as a dashed line.

When signal frequencies are lost, distortion results. It is possible to remove the signal entirely, leaving only background, with a large enough filter window. How large a window can be used without significant distortion? The window should not be larger than the *smallest* peak half-width present in the signal. This rule ensures that the narrowest peaks (with the highest component frequencies) will not be clipped off by the moving-average filter.

The moving-average filter can be compared to the other smoothers discussed above by transformation of the filter impulse function to the frequency domain, which produces the filter transfer functions shown in Figure 15. The general shape of the moving-average filter transfer function is similar to those designer transfer functions discussed above, but with some important differences. One difference is the presence of negative regions in the moving-average filter transfer function. These regions will produce a filtered, frequency-domain representation with *negative* intensities at some frequencies after multiplication of the filter transfer function by the frequency-domain representation of the noisy signal. These negative intensities in the frequency domain lead to phase errors [4] in the filtered output, so that a decrease in the number of negative regions in the transfer function is desirable. There are also a number of humps in the transfer function, where higher frequencies can "leak" through to the filtered output. Again, a decrease in the number of these positive areas is desirable, if possible.

Problem 5. One way to create a designer transfer function is to convolve the smoother with itself several times before applying the new smoother produced by this convolution operation to the noisy data. Sketch the transfer function that

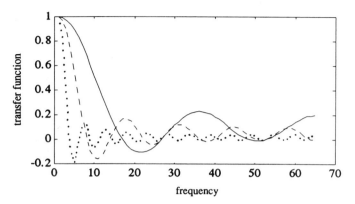

Figure 15 Transfer functions for FIR filters as a function of window size. Transfer functions for simple moving-average filters with windows of 5 (solid line), 10 (dashed line), and 25 (dotted line) are shown.

would result if the smoother described in Figure 15 were convolved with itself four times. (*Hint*: Repeated convolution leads to a Gaussian shape as a consequence of the central limit theorem. See reference 4 for details.)

Polynomial Moving-Average (Savitsky–Golay) Filters

From Figures 13 and 14, it is clear that a moving-average filter provides significant noise reduction. However, signal distortion also occurs, as can be seen in Figure 14, which compares the smoothed signal obtained from application of a moving-average filter using a 10-point window and the true signal $s(t)$. The distortion produced by the filtering process is apparent. To reduce this distortion while retaining noise attenuation, simple polynomials can be fitted to data in a manner analogous to the moving-average filter discussed above. Polynomial moving-average filtering involves fitting a polynomial to a set of noisy data within a window of points, as shown in Figure 16. Using higher-order polynomials should allow fitting of data that change rapidly within the filter window. As long as the noise continues to change more rapidly than the data, good noise rejection will be possible. Any set of polynomials could be used to define a polynomial filter [11]. Selecting the set is analogous to choosing a frequency-domain designer transfer function. For example, Chebychev polynomials can be used to filter data in the time domain; the transfer function of the Chebychev polynomial moving-average filter was given in Figure 7. One especially convenient set of filter polynomials is based on power functions. Savitsky and Golay first introduced the use of moving-average filtering with orthogonal polynomials based on power functions, and many texts in chemistry now refer to the method as Savitsky–Golay filtering [10,12–14]. As with other FIR filters, the polynomial is convolved

SIGNAL PROCESSING AND DATA ENHANCEMENT

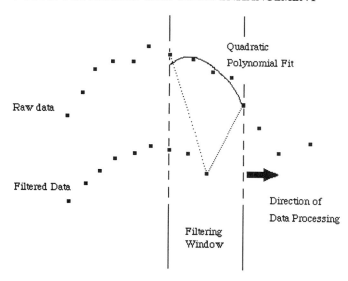

Figure 16 Polynomial least squares filtering. A quadratic, five-point polynomial filter is shown.

with the noisy time-domain data to produce an enhanced representation of the true signal. Generation of the filter polynomial from the power series is simple [15], and many authors provide listings of the polynomials [12–14].

By selecting the order of the convolving polynomial, the user is able to alter the filtering process. With low-order polynomials, more noise is filtered, but at the cost of increased signal distortion. In fact, the moving-average filter is just a very low order (order 1) polynomial moving-average filter. Increasing the size of the window used in the filtering also decreases noise, as with the moving-average filter, but at the cost of significant signal distortion [4]. Filtering with low-order polynomial functions and large filter windows is called *strong filtering*, and it is used when obtaining a high S/N ratio is more important than preserving signal shape. When filtering is done with small filter windows and higher-order polynomials, less noise is removed because the higher order polynomials will describe a rapidly varying signal (and, of course, noise) better than low-order polynomials, but less signal distortion will be caused in filtering. Filtering with high-order polynomials is known as *weak filtering* and is mainly of use when preservation of signal shape, not S/N ratio, is most important [16,17]. Figure 17 demonstrates the results of weak and strong filtering on the noisy signal shown in Figure 1.

Problem 6. Window size and polynomial order both play a part in the noise reduction process for a polynomial moving-average filter. For a given polynomial

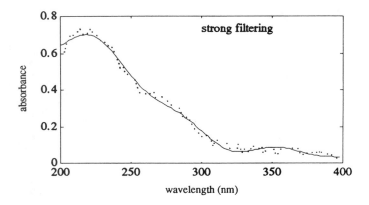

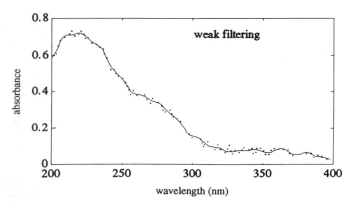

Figure 17 Weak and strong filtering of noisy data using quadratic polynomials. Weak filtering was done with a 20-point smoothing window, while strong filtering was done with a five-point smoothing window.

order, transfer functions are plotted for three window sizes in Figure 15. Which provides the strongest filtering, and which the weakest filtering?

8.5 RESOLUTION ENHANCEMENT IN THE TIME AND FREQUENCY DOMAINS

Deconvolution

It sometimes happens that the noise corrupting a signal arises from an instrumental artifact or from an uninteresting physical process. For example, the

SIGNAL PROCESSING AND DATA ENHANCEMENT

broadening of a spectral line by the spectrometer slit function, or the distortion of a current versus time relation by diffusion can be considered a kind of "noise" on the desired (unaffected) signal. These "noise" sources directly overlap the signal in both the frequency and time domains, so that simple attenuation of the undesired response is not possible by smoothing. Instead, we take advantage of the fact that we know something about the shape of the perturbing effect. Since the effect of these noise sources is a convolution of the noise source $n(t)$ with the desired (true) signal $s(t)$ to produce a distorted, observed signal $x(t)$, we might be able to enhance the observed signal by *deconvolution* of the known noise source from the observed signal. Deconvolution is easiest in the frequency domain, where the operation is simply the division of the transform of the observed signal by the transform of the perturbing function $n(t)$,

$$S(f) = X(f) / N(f) \tag{20}$$

Unfortunately, the division introduces some difficulties. At frequencies where $X(f)$ and $N(f)$ are both small, their quotient need not be, and it often is not. The increase in the amplitude of frequencies not carrying much signal information implies that as the convoluted noise source is removed, other noise must be introduced, and the S/N ratio of the deconvoluted data generally decreases, as shown in the sequence in Figure 18. For this reason, deconvolution of data is generally done in concert with smoothing and apodization [18].

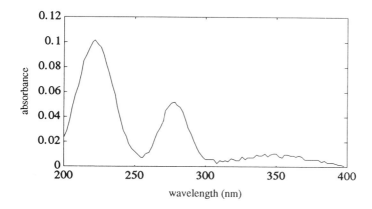

Figure 18 Deconvoluted data from Figure 1. A triangular slit function shape was assumed for the deconvolution, and the noisy data were smoothed prior to the deconvolution step.

Derivative Methods

Often, even the approximate nature of the interfering process is unknown, and deconvolution of the process from the observed signal is not possible. In these cases, other methods may be used to enhance the resolution of the data. The simplest of these methods rely on derivatives to sharpen the broadened features. If numerical derivatives are used, the functional form need not be known for either the observed signal or the broadening effects. Another advantage of using derivatives to enhance resolution is that higher order derivatives can be used to resolve very strongly overlapped responses. Figure 19 shows a first-derivative spectrum obtained by a simple differencing approach. This spectrum was produced after smoothing the data, because the differencing operation strongly increases the higher frequencies, which contain mostly noise.

Another way to enhance resolution while minimizing noise involves the use of sets of polynomials based on derivatives in the polynomial smoothing methods discussed above. In this way, approximate derivatives can be taken in the time domain, while the data are simultaneously smoothed to decrease noise [9,12,15]. One advantage of this method is that selection of the appropriate window for smoothing, the order of the polynomial, and the order of the derivative sets, in effect, the shape of the transfer function directly. It is very convenient to couple the operations of smoothing and differentiation, by use of the moving-average polynomial filter discussed earlier. This coupling permits weak or strong filtering to be done while data are converted to first or higher derivative representations. The same considerations of polynomial order and window size apply as with ordinary polynomial filtering. In this case, since the

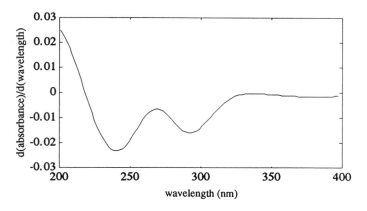

Figure 19 First derivative of noisy spectrum in Figure 1 obtained by simple differencing after smoothing.

SIGNAL PROCESSING AND DATA ENHANCEMENT

scales are so different from the raw data and the derivative, the unfiltered ends dominate the result, and they must be removed for display. Figure 20 shows a first-derivative spectrum generated directly from the noisy data in Figure 1 by a quadratic, first-derivative filter using an 11-point window. Some more noise is seen in the derivative, as a result of the relatively small window size chosen to avoid distortion of the differentiated peaks. An advantage of this approach is speed. The filtering and differentiation can be done online.

Derivatives can be taken directly, in the time domain, or they can be taken in the frequency domain. The derivative property of the Fourier transform, given in Table 1, indicates how the two domains define the differentiation operation. According to the derivative property, taking a derivative in the Fourier domain is simple: The transform of the original data is multiplied by a linear function. The result is the transform of the first derivative of the data, a fact that is confirmed by transformation of the result back into the time domain. Figure 21 shows the result of this operation.

The use of derivative methods is subject to the same noise enhancement problem seen in deconvolution. Higher order derivatives further increase the noise in the resolution-enhanced spectrum. One reason for the noise enhancement is the derivative operation itself. The multiplication of the frequency-domain representation of the data by a linearly increasing window function, required to obtain differentiation in the frequency domain, acts as a high-pass filter, emphasizing high-frequency components of the signal while decreasing low-frequency components. Noise will therefore tend to be enhanced in the operation of differentiation.

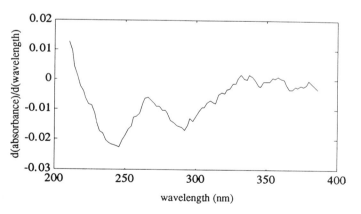

Figure 20 First-derivative resolution of overlapped responses in Figure 1. Smoothing was done first to avoid significant noise effects in the derivative.

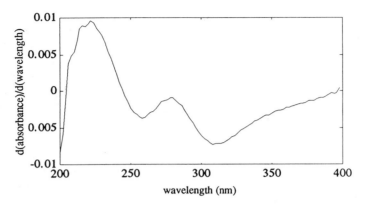

Figure 21 Frequency-domain differentiation of data.

To minimize the problem of noise enhancement while taking derivatives of data, it is possible to alter the transfer function of the differentiation operation. One way is to create a designer transfer function for differentiation, which produces an *approximate* derivative from the Fourier domain differentiation operation [5,6]. Thus, both differentiation and smoothing can be built into the transfer function, as shown in Figure 22.

Because the goal is resolution enhancement, and not generation of the exact derivative, this ad hoc method can be adjusted to suit the data, enhancing resolution while simultaneously minimizing the noise enhancement. Close examination of Figure 21, which was produced by application of this approximate

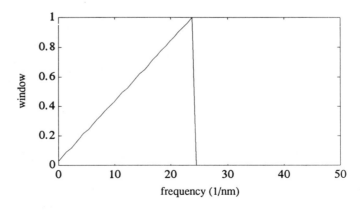

Figure 22 Derivative smoothing transfer function.

derivative, shows some differences from Figures 19 and 20. All three spectra are derivative spectra, but all are approximate derivatives because of the need to remove noise. Thus, small differences in the three approximate routes to the same product lead to small differences in the derivative spectra.

8.6 SUMMARY

The dual nature of the frequency and time domains leads to sets of complementary methods for data processing. Frequency-domain methods often are simpler for smoothing and deconvolution operations, while time-domain methods offer advantages in filtering and differentiation of data.

The interested reader will discover that frequency-domain analysis does not stop with Fourier-domain representations; indeed, the Hadamard transform, a transform very much like the Fourier transform but based on binary sequences rather than sines and cosines, can be used for signal processing [19]. Other transform methods, including those based on the z transform and Hilbert transform also offer many attractive features [3]. The z transform in particular permits considerably more to be done with filtering and smoothing of sampled data. There is a considerable body of literature concerned with z-transform approaches to signal processing [3,20].

Time-domain signal processing methods also go well beyond those for which simple frequency-domain analogs can be described. Infinite impulse response (IIR) digital filters [3,21] and recursive digital filters [21,22] are two time-domain methods with uses in chemistry. Resolution enhancement by time-domain methods is also possible though beyond the scope of discussion in this chapter. One method, based on maximization of information entropy, is now routinely used for resolution enhancement of noisy spectral data, including two-dimensional nuclear magnetic resonance [23].

SOLUTIONS TO PROBLEMS

1. Least-squares fitting of the true signal x to the noisy signal y, where $y = ax + e$, gives $a = 1.681$. The noise e is given by the residuals $(y - \hat{y})$. The peak S/N ratio is given by the maximum signal ax divided by σ_{noise}, which is $1681/76 = 22$.
2. The discrete convolution of r and s is 0 0 4 2 0 −2 0 0 4 6 2 0 0 2 0 0 −2 0 0 2 2 2 0 0. The discrete convolution of diminished r and s, done as suggested in the exercise, is 4 2 0 −4 −2 0 0 4 6 6 0 −2 0 0 2 2 2 2.
3. The correction is necessary to keep the total power constant. Negative frequencies carry half the total power, and their loss would decrease the signal intensity if the correction were not made.

4. The preprocessing step is needed to ensure that both ends of the data have the same value. The cyclic nature of the convolution means that any difference between ends appears as a "step function" in the data. The transform of a step function is a (sin x)/x ("sinc") function, which appears as spurious oscillations in the transformed data.
5. Self-convolution produces a Gaussian shape. The result of four self-convolutions is a transfer function that mimics a half-Gaussian. Further, each convolution reduces the half-width, so the self-convolved smoother has a much narrower passband than the original smoother.
6. The strongest filtering is with the largest window, since this gives the maximum S/N enhancement. The weakest filtering is with the smallest filter window.

REFERENCES

1. R. N. Bracewell, *The Fourier Transform and Its Applications*, 2nd ed., Wiley, New York. (1972).
2. W. H. Press, B. P. Flannery, S. A. Teukolsky, and W. T. Vettering, *Numerical Recipes: The Art of Scientific Computing*, Cambridge Univ. Press, Cambridge. (1986).
3. N. K. Bose, *Digital Filters: Theory and Applications*, Elsevier, New York. (1985).
4. P. D. Wilson and T. H. Edwards, *Appl. Spectrosc. Rev.*, *12*: 1 (1976).
5. G. Horlick, *Anal. Chem.*, *44*: 943 (1972).
6. K. R. Betty and G. Horlick, *Appl. Spectrosc.*, *30*: 23 (1976).
7. J. F. Kaiser and W. A. Reed, *Rev. Sci. Instrum.*, *48*: 1447 (1977).
8. S. A. Dyer and D. S. Hardin, *Appl. Spectrosc.*, *39*: 655 (1985).
9. S. E. Bialkowski, *Anal. Chem.*, *60*: 355A (1988).
10. C. G. Enke and T. A. Nieman, *Anal. Chem.*, *48*: 705A (1976).
11. P. D. Willson and S. R. Polo, *J. Opt. Soc. Am.*, *71*: 599 (1981).
12. A. Savitsky and M. J. E. Golay, *Anal. Chem.*, *36*: 1627 (1964).
13. J. Steiner, Y. Termonia, and J. Deltour, *Anal. Chem.*, *44*: 1906 (1972).
14. H. H. Madden, *Anal. Chem.*, *50*: 1383 (1978).
15. S. E. Bialkowski, *Anal. Chem.*, *61*: 1308 (1989).
16. M. U. A. Bromba, *Anal. Chem.*, *53*: 1583 (1981).
17. M. U. A. Bromba and H. Ziegler, *Anal. Chem.*, *56*: 2052 (1984).
18. P. J. Statham, *Anal. Chem.*, *49*: 2149 (1977).
19. J. Zupan, *Algorithms for Chemists*, Wiley-Interscience, New York, Chapter 5. (1989).
20. D. G. Childers, ed., *Modern Spectrum Analysis*, IEEE Press, New York. (1978).
21. S. E. Bialkowski, *Anal. Chem.*, *60*: 403A (1988).
22. S. D. Brown, *Anal. Chim. Acta*, *181*: 1 (1986).
23. S. Haykin, ed., *Nonlinear Methods of Spectral Analysis*, 2nd ed., Springer-Verlag, New York. (1983).

FURTHER READING

Brigham, E. O., *The Fast Fourier Transform*, Prentice-Hall, Englewood Cliffs, N.J. (1974).

IEEE, *Programs for Digital Signal Processing*, IEEE Press, New York. (1979).

Jenkins, G. M., and D. G. Watts, *Spectral Analysis and Its Applications*, Holden Day, San Francisco. (1968).

Nussbaumer, H. J., *Fast Fourier Transform and Convolution Algorithms*, Springer-Verlag, New York. (1982).

Oppenheim, A. V., and R. W. Schafer, *Digital Signal Processing*, Prentice-Hall, Englewood Cliffs, N.J. (1975).

Parks, T. W., and C. S. Burrus, *Digital Filter Design*, Wiley, New York. (1987).

Rabiner, L. R., and B. Gold, *Theory and Applications of Digital Signal Processing*, Prentice-Hall, Englewood Cliffs, N.J. (1975).

9
The Role of the Microcomputer

K. L. Ratzlaff *The University of Kansas, Lawrence, Kansas*

Eugene H. Ratzlaff *IBM Research Division, T. J. Watson Research Center, Yorktown Heights, New York*

Over the past decade small computers have enabled the movement of computation from the large time-shared mainframe computer to the small self-contained laboratory computer. The revolution in microelectronics has driven this change, at its center bringing a central processing unit (CPU) [1] in a single integrated circuit.

This revolution has two consequential ramifications for the area of chemometrics. First, the availability of low-cost, small computers provides a choice of environment for computing. Second, and most significant, these computers can support both the acquisition and processing of the data, thereby enabling very powerful systems that can close the experiment/data processing loop.

In this chapter, we describe the microprocessor and the microcomputer as they relate to chemometric operations. Issues of performance, programming, interfacing, and operation are central to understanding where these tools belong.

9.1 THE MICROPROCESSOR

The microprocessor came to the computing scene in the middle 1970s with more promise than was generally realized [2, 3]. At that time, small computers such as PDP-8s and Novas were common in scientific laboratories, probably most common in chemist's laboratories. However, their role was usually limited to instrumental data acquisition and control. The limitation was imposed, not necessarily by the lack of power in the processor, but by three handicaps:

1. Memory size—memory of 4K to 8K words of 12-bit memory was common, and 64K-word memory was considered large.
2. Software availability—writing one's own software in assembly language or, at best, in FORTRAN was the norm.
3. Online storage capacity and speed—paper tape and cassette tape were common, while hard and floppy disks were slow, bulky, and expensive.

Since the first microprocessor, there has been an increase in processor system capability exceeding two orders of magnitude, accompanied by a decrease in the cost of both memory and hard disk storage of three orders of magnitude. Clearly, the computer based on the microprocessor is capable of handling many of the problems that a few years ago were performed exclusively on the time-shared mainframe computer.

The reasons for using the small computer rather than the time-shared machine fall generally into two classes: (1) the advantage of having complete control over the computer and (2) the need to acquire data directly from an experiment while maintaining intelligent control of it.

The Online Computer

While this volume deals largely with the processing, handling, and interpretation of data, this chapter also examines the acquisition of that data, in part a historical discussion.

For the first three decades of digital computing, the economies of computers mandated that computers be shared. No single task could be allowed to dominate the priority structure, and consequently computers were not used to acquire the data. Experimental results typically were plotted on a strip-chart recorder or displayed on a panel meter from which they had to be transferred to a medium that could be read by the computer. The term *offline* [4] is used to describe this sequence. The operator, as in the diagram of Figure 1, is a link in the movement of information from the experiment to the computer and back again to the experiment. Although the computer's task may be completed rapidly, the entire loop is slow and error-prone.

The demand for computers that could take on the task of data acquisition, driven initially by the U.S. space program, brought into production the small computers that could be dedicated to the demands of a scientific experiment. The data are acquired directly by the computer and can be stored, processed, and used to control variables in the experiment.

THE ROLE OF THE MICROCOMPUTER 273

Figure 1 Schematic representation of offline computer operation. Note that the experimenter is part of the data path. (Reprinted from reference 1 by permission of John Wiley & Sons, copyright © 1987.)

The term **online** describes this process because the computer communicates directly with the experiment as outlined in Figure 2. Consequently the loop from the computer, where the data are processed, back to the experiment is reduced to the submillisecond time scale. The user/operator interacts directly with both the computer and the experiment, remaining fully in control but out of the feedback loop.

Very significant advantages can be associated with online operation. First, data can be acquired both with high integrity and at relatively high rates (approaching 1 MHz even without specialized tools); in the reverse direction, the results of sophisticated processing can be used for control of the experiment while the experiment is in progress, that is, in *real time*. Second, the user/operator has full control of the software and algorithms that are applied to the data.

The discussion of modes of computer operation would not be complete without mention of the instrument with an embedded (onboard) microcomputer. This mode shares with online operation the direct transfer of information between experiment and computer but lacks the versatility of direct interaction of the user with both the computer and the experimental variables: as Figure 3 suggests, all operations pass through the computer. This mode is known as the *inline* mode [1,3].

For purposes of this discussion, the distinguishing feature of inline operation is that the computer cannot generally be programmed by the user; although it is capable of very sophisticated chemometric operations, the operations are limited to those that were installed by the instrument vendor, often with very inadequate description.

Inline operation is now the rule for commercial chemical instrumentation, but when further processing of the data is required, an additional online computer is required.

Clearly, the ideal situation will be to have the power and performance (terms to be discussed later) of a large mainframe computer in an online labo-

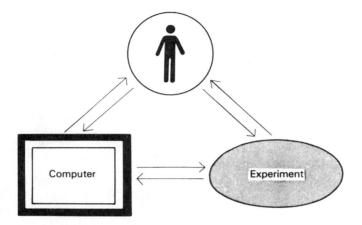

Figure 2 Schematic representation of online computer operation. The experimenter is outside the data path, interacting with both the computer and the experiment. (Reprinted from reference 1 by permission of John Wiley & Sons, copyright © 1987.)

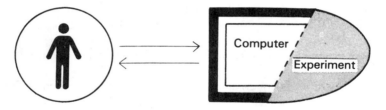

Figure 3 Schematic representation of inline computer operation. The experiment and the computer are now indistinguishable. (Reprinted from reference 1 by permission of John Wiley & Sons, copyright © 1987.)

ratory computer. This is a realistic expectation for most environments, as will be shown later.

Computer Description

The small computer can most usefully be described from an *operational* point of view, considering the components that are commonly found or are available in the small computer. The block diagram of Figure 4 shows the fundamental components that will be described individually, not so much from a technical point of view, but in consideration of how those technologies affect the chemometric environment.

THE ROLE OF THE MICROCOMPUTER

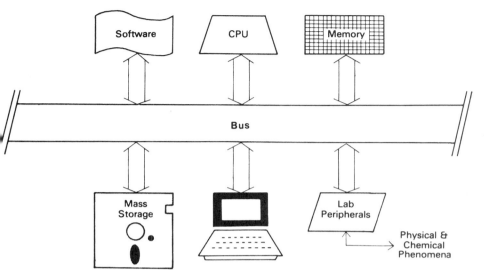

Figure 4 Block diagram of the major components of a small computer suited for online operation. (Reprinted from reference 1 by permission of John Wiley & Sons, copyright © 1987.)

Central Processing Unit

The central processing unit (CPU) is the component that executes (processes) individual instructions. *Microprocessors* are simply CPUs that are housed in a single integrated circuit (IC), and one such IC is shown in Figure 5. Each CPU design executes an instruction set that is different from the instruction set of another CPU; consequently, the actual code that is produced for one CPU will not execute on another.

CPUs execute instructions in steps that are driven by a *system clock* that governs the rate of operation. This factor and others that dictate system performance will be discussed in a following section.

Memory

The computer's memory holds both the instruction sequence and the data during computer operation. Important features to be considered include the size (or capacity) of the memory array and its speed.

The cost of semiconductor memory (sometimes incorrectly called *core* [1]) has become a much smaller part of the expense of a small computer. Consequently, the very large memory capacity that would be useful for many chemometric applications should, in principle, be practical. However, in many cases

Figure 5 The die of an Intel 80486 microprocessor. The CPU contains the equivalent of 1 million transistors. (Photo courtesy of Intel Corporation.)

the software, both the operating system and the compiler, limits the amount of memory that can be used. This limitation is rapidly disappearing.

The *speed* of the memory can often become a limiting factor in the performance of a small computer. When the program code calls for the CPU to read from or write to memory, the memory must be able to respond within a part of the clock cycle of the system clock. If memory is used that cannot operate on that time scale, a *wait state* must be added; the CPU is forced to wait for an additional system-clock cycle while the memory circuit settles. Typically, a single wait state will decrease system performance by 25–30%.

One technique, borrowed from mainframe computers, that is used to avoid the expense of providing the new, fast CPUs with very fast memory is to use *cache memory*. Cache is a relatively small amount of memory that can respond within the cycle time of the CPU when the main memory of the computer cannot. An entire bank of currently used memory is moved into cache and executed from there [5]. Dedicated cache memory manager circuitry controls

THE ROLE OF THE MICROCOMPUTER

these operations, and in several new CPU designs both the cache and the cache management are integrated into the microprocessor. When considering real-time operation, it should be noted that cache is a two-edged sword: whereas the *average* memory access may improve, the *worst case* will be degraded because whenever a memory access goes outside the region contained in the cache, an entire block must be moved [5].

User Input/Output (I/O)

The user receives most of the output via a video screen, and standard schemes provide increasing resolution and color capability; graphics capability has gone from being an extra-cost enhancement to becoming a necessary component. For those computers that are marketed as general-purpose microcomputers, a resolution of 640 × 480 in 16 colors is currently typical.

Multiple processes or programs may be viewed simultaneously by segmenting the video screen into tiled or overlapping segments called *windows*, one window for each process. Engineering and scientific workstations and video displays for users of windows and *graphical user interfaces* (GUIs) often provide resolution of 1024 × 768 to 1280 × 1024 or more.

While textual information is entered via keyboard, the efficiency of user input is enhanced by the addition of *pointing devices*, input devices amenable to the manipulation of graphical information. Even textual information can be manipulated graphically using icons—images that represent information or processes. Information represented by an icon is further manipulated in windows. Dominant among pointing devices is the *mouse*, a very natural link between the user and an image; trackballs and touch screens may be suitable for certain applications. The acronym WIMP (Windows-Icons-Mice-Pointing devices) represents the overall approach.

Graphical output is equally important. Normally a *plotter* is employed, particularly for multicolor drawings, but a printer can often provide equal resolution and greater convenience. A desktop laser printer may be faster and provide a sharper image, thereby becoming the graphical output instrument of choice.

Communications Input/Output

In this chapter, *communications* refers to the capability of transmitting information between computers. Communication takes place at several levels, starting with one computer acting as a terminal.

For simple communications, the computer emulates a computer terminal, communicating via the serial port [6]. The software for this task is often enhanced with capabilities for sending and receiving files. The latter capability requires protocols for handling errors that may occur during transmission, and the same protocol must be implemented on both ends of the communications link. There is no single accepted standard for the protocol; two of the more common protocols are Kermit and Modem-7.

A local-area network (LAN) with high-speed transmission between a group of small computers facilitates the sharing of databases, files, mass storage, and resources such as printers, plotters, and modems. To an individual user, the system operates as though resources on other computers were actually a part of the local computer. Electronic mail may be part of these LANs. Local-area networks in a laboratory environment may not yet have reached their highest potential, since the online demands made of a computer necessitate greater reliability, flexibility, and performance than is satisfactory for the office environment [7].

At a much more sophisticated level lies the possibility of a network in which not only is information shared as easily as a file is read or stored, but actual computational tasks can be shared between CPUs.

A wide-area network (WAN) typically performs different functions. At relatively low speed and cost, store-and-forward networks are used to transmit mail and files; BITNET is such a network, linking universities and nonprofit research institutions, primarily in North America and western Europe. Each message or file is routed by sending it from one BITNET node to the next. High-speed WANs provide connectivity between users and remote computers; for example, NSFNET [8] connects users at universities throughout the United States with supercomputer sites at rates expected to reach megabits per second, sufficient for supporting local graphics workstations.

Laboratory Interfaces

Of particular importance in chemical applications is the ability to acquire information directly from the experiment and control the experiment with the computer. Data acquisition modules added to the small computer system provide that capability, and these modules and their applications will be described in Section 9.4.

Mass Storage

The capacity for storing massive amounts of data online was once a characteristic of large, shared computers. However, the microcomputer revolution has driven this technology as well. The ubiquitous floppy disk has developed from an 8-in.-diameter medium storing 128 Kbytes in the mid-1970s to a 3½-in.-diameter unit storing well over 1 Mbyte. Rigid magnetic disks, once physically larger than current computers for 5-Mbyte capacity now store over 700-Mbytes in the space occupied by a floppy disk drive. The time required to access a block of data on the disk is also plummeting.

The newer optical disks capable of storing gigabytes on a surface are already finding their way into applications where large amounts of data must be stored permanently. Most optical disk technologies, some based on compact disk (CD) technology from the audio world, are ready-only—the information is written at the factory and cannot be modified by the user. Large databases on read-only optical disks are readily searched by a small computer, and several databases for scientists are available on optical disks; *ChemLink* (Chemron Inc., 3038 Orchard Hill, San Antonio, TX 78230) is a database with search software containing the information from the catalogs of 31 chemical suppliers [9]; *Aldrichem Data Search* (Aldrich Chemical Co., P.O. Box 355, Milwaukee, WI 53201) is a compendium of reference data, physical data, and structural data on 50,000 chemicals; the *Sigma-Aldrich Material Safety Data Sheets* (Alrich Chemical Co.) provide 33,000 material safety data sheets searchable by catalog number, CAS number, name, or formula [10]; *CD-Chrom* (Preston Publications) gives access to 70,000 chromatography literature abstracts; *PC-PDF* (JCPBS—International Centre for Diffraction Data, 1601 Park Lane, Swarthmore, PA 19081) contains the entire PDF-2 powder diffraction database and index files.

WORM (write-once-read-many) drives are ideal for regulated industries such as pharmaceuticals where huge volumes of test data must be stored permanently. Information can be written locally rather than at the factory, but, compared with magnetic media, it has a nearly infinite lifetime.

Newly available magneto-optical (MO) drives combine the benefits of the rigid disk's storage capacity and rewritability with the floppy disk's media portability, memory expansion, and offline storage capacity. Magneto-optical drives will undoubtedly find wide use in many microcomputers and workstations as reliability is established and access time improves.

The Computer's Expansion Bus

The internal expansion bus, containing a group of equivalent slots, is singled out for comment even though it is a nearly passive component of the computer. (There are other buses in the computer that should not be confused with the computer bus—for example, the instrument interface bus, described in Section 9.4). The capacity for accepting add-on components provides necessary versatility to laboratory computers. Computers that lack a bus (e.g., Commodore-64 and many Macintosh computers) cannot readily accept computational enhancements or tools for data acquisition and control.

The bus design can limit the performance of a computer in two ways. First, the physical layout of the electrical conductors affects the maximum rate at which data can be transmitted reliably over the bus. Second, and more important, the bus must contain sufficiently many conductors (that is, it must be "wide" enough) to carry all the signals without suffering the penalty of using the same line to carry more than one signal (multiplexing).

Software

Mentioned here simply for completeness, it should be pointed out that no computer can operate usefully without software; a fuller discussion will be given in Section 9.3.

System Performance

Everyone needs "performance," but seemingly nobody can get enough. Tasks such as word processing should be mundane to a small computer, and a low-end machine should suffice. However, one soon learns that word processing with a high-end workstation is so much more convenient that it is difficult not to "need" that performance. The purpose of this section is to understand what performance is and the factors that go into building it.

Factors Related to Performance

Computers are often compared with respect to their performance with greatly varying results. It is not easy to define or measure performance, and even if it were possible, a long list of contributing factors would have to be considered. To that end, this section will attempt to review those factors, considering ways that could be used to obtain more performance from a computer.

Clock Speed. The simplest way to increase the performance of a computer might be to increase the clock speed, and many otherwise identical computers differ in clock speed which can be as great as 40 MHz at this writing. The ultimate limit is physical; any CPU will fail if the clock speed is pushed beyond its limits. Even if the CPU is capable of operating at high clock speeds, the memory and peripheral devices might not be; slow memory and peripherals might be accommodated by adding wait states with the penalty of reduced computational throughput. However, if the support circuitry does not place constraints on the CPU, the throughput for a given CPU will be directly proportional to the clock speed.

Word Size. A processor is designed to fetch, manipulate, or store a single word of instruction or data in most of its operations. All other factors being equal, performance should improve when a CPU with a larger word size is chosen. In software, the definition of a word is 16 bits (or 2 bytes), but in the hardware it is the number of bits that are processed in parallel. The word size of very early microprocessors (e.g., the Intel 4004) was only 4 bits, and the first microprocessor that became the basis of a personal computer (the Intel 8080) [2] had a word size of 8 bits. Personal computers of today are dominated by 16-bit (e.g., the Intel 8086) and 32-bit (e.g., the Intel 80386/80486 and Motorola 68030) microprocessors. It should be remembered that the word size does not affect the precision of the computation that can be completed by the computer.

Although these definitions are commonly accepted, it should be noted that CPUs do not operate purely with a single word size. Eight-bit CPUs often have some 16-bit instructions, for example.

CPU Architecture. The CPU designer's task is to develop a CPU that will handle computing tasks as efficiently as possible within the constraints of the integrated circuits that are to be manufactured. At this writing, two classes of CPUs are competing to be the preferred architecture: Reduced Instruction Set Computers (RISC) and Complex Instruction Set Computers (CISC). The CISC CPUs, which have dominated the computer designs of the past, have a large number of instructions that execute in one to many clock cycles. By contrast, the objective of the RISC design is to execute all instructions in a single clock cycle; to accomplish that, the designers must be satisfied with a smaller repertoire of instructions, but the clock rates may be much higher.

Although such factors as memory speeds and the body of programming experience may weigh heavily in favor of CISC technology, the RISC processors tend to be capable of greater raw computing power. The addition of special functions for computation, graphics, or text handling to a RISC processor gives it a large advantage for specialized applications.

Since most RISC instructions are executed in a single clock cycle, RISC processors can potentially be used in parallel with much less programming complication than would be required to coordinate parallel CISC processors. When the same processors are used, parallel operation provides the potential multiplicative improvement compared with the linear improvement of increasing the clock speed.

Supercomputers, such as the Cray X-MP, achieve their high level of performance with another innovation, vector architectures, which enable them to process numbers in a "pipeline," where each step of a multicycle instruction is executed in sequenced stages. All stages operate in parallel, but the multicycle instruction operates on each datum as it "flows" through the pipeline. The application software must be designed to take advantage of this feature; the pipeline is not efficiently used if it must change its operation for each datum. The program must be organized so that an entire data vector is subjected to a single operation before the next operation is begun (vectorized) in order to obtain the full potential of the computer.

Accelerators. One of the ways in which performance of a desktop computer can be enhanced is by adding components for special operations, particularly for floating-point computation. Of several tools that are available, coprocessors enhance the instruction set by performing specialized but powerful operations, array processors operate on data arrays using pipeline techniques, and CPU accelerators can substitute more powerful CPUs without changing the rest of the computer.

Floating-point math coprocessors are available for the leading microprocessors: 6888x coprocessors for the Motorola 680x0 series, and 80x87 coprocessors for the Intel 80x86 series. They often speed execution of computation-intensive software by an order of magnitude. Compiled software typically searches for and uses the devices if they are installed. Other coprocessors are used to take over graphics, character string search, or input/output tasks.

An array processor makes use of a pipelined architecture for very high throughput of mathematical operations. As such, the benefits of an array processor are found only when the same operation is applied to a large array of data. One frequently used figure of merit is the time required for a single-precision 1024-point fast Fourier transform (FFT); the Sky Warrior-II array processor (Sky Computers, Inc., Foot of John St., Lowell, MA 01852) executes it in 2.25 ms (nearly three orders of magnitude faster than a 16-MHz 80386 computer with an 80387 coprocessor). Special compilers or libraries are usually required to use array processors.

A CPU accelerator is special-purpose hardware that is used to speed particular processes. The term can refer to add-in boards that allow a CPU upgrade without replacement of the entire computer; this approach is usually a compromise because the computer bus may not have the necessary bandwidth or the full address size, or the memory may not be sufficiently fast. Other accelerators handle specialized tasks that would be handled by the CPU in the standard computer: graphics operations (converting vectors to raster patterns, filling an area, performing two- or three-dimensional transformations, or rendering a three-dimensional solid), cache memory handling for disk drives, character-string searching for text operations, and so on. These accelerators will increasingly be adapted RISC CPUs.

Software. The skill with which software is written dramatically affects performance, and this applies both to how the application code is written and to the quality of the compiler. When compilers are compared (benchmarked) using the same program compiled by each, it is not unusual to find a variation in execution time exceeding a factor of 2, a performance component that may equal processor speed in importance.

Some Current Comparisons

The question of how to compare the performance of various computers is far from settled. The only definitive comparison would be to test one's own application program on each computer, if the specific applications were already well known.

THE ROLE OF THE MICROCOMPUTER

Several numerical standards are used frequently: MIPS (millions of instructions per second); MFLOPS (millions of floating-point operations per second); Dhrystones (execution speed of a standard task); and several well-known FORTRAN floating-point programs used as benchmarks: the Whetstone, the Linpack, and Livermore loops. (Because the latter is very popular, it is commonly alleged that some compilers demonstrate improved benchmark performance by searching for Livermore loops and computing them with optimized code.)

A new set of 10 programs call the SPEC benchmark is emerging from major vendors in the Unix scientific and engineering market. The SPEC benchmark should prove useful for intercomparisons of technical workstations where recently introduced desktop and deskside computers, such as the IBM RS/6000, outperform the Cray X-MP in scalar MFLOPS.

Any attempt to compare computers should be greeted with some caution, but the data presented in Table 1 do give a very rough comparison [11].

Directions in Microcomputing

It can be useful, even if risky, to speculate on the directions that desktop computers might be taking in the future. The two areas in which the greatest changes might be found are in performance and connectivity.

Performance

Changes in computer performance are largely evolutionary, seldom revolutionary, no matter what the pundits say, but that evolution occurs at a rapid rate. In the mid-1970s, as the microcomputer began to be employed as a desktop computer, it would have been difficult for even the most forward-looking com-

Table 1. A Rough Benchmark Comparison of Current Computers

Computer	MFLOPS
Cray X-MP (vector mode)	24
Cray X-MP (scalar mode, program not vector-optimized)	6
DECstation 3100	1.8
Sun 4/260	0.86
80386/3167 (with Weitek coprocessor, 25 MHz)	0.5
80386/80387 (25 MHz)	0.25
MicroVAX 2	0.13
Macintosh IIx, SE/30	0.03
8088/8087 (IBM PC/XT, 8 MHz)	0.01

Source: Cameron [11].

puter user to foresee the capability of today's desktop computers. Since the introduction of the IBM PC, there has been a 50-fold increase in raw computing ability. A rule of thumb is sometimes invoked that performance will double every year for the next decade, at approximately the same cost; it has held over the past decade. Memory, graphics, and disk storage costs drop dramatically as the technology evolves.

An outline of the expected evolution of the 80x86 family for the coming decade projects 4 million transistors on an IC in the 80586 in 1993, and 16 million in 1996 [12]. A clock rate of 250 MHz is being foretold.

One of the other routes by which the higher performance is being achieved is to combine several functions in a single chip, reducing delays and permitting pipelined and parallel execution. The Intel 80860 is a RISC chip capable of 50-MHz operation with a 17-MFLOP floating-point unit on the chip and a three-dimensional graphics processor on the chip [13].

Increasing parallelism has always played a role in improving computer throughput by offloading many CPU tasks to intelligent controllers for disk, communications, keyboard, timer, graphics, floating point, and so on. This parallel distribution of specific responsibilities will continue; however, CPU parallelism will also inevitably have an impact on the desktop workstation. The *transputer* is an emerging example of a microprocessor designed expressly for CPU parallelism. The transputer combines RISC-like features, extensive inter-CPU communications capabilities, and multiprocessing software support through parallel compilers and an explicitly parallel language called OCCAM. Development systems and multiprocessing accelerators with capacities for several transputers are available for desktop hosts and workstations.

Even with these spectacular advances, the appetite for computing in the laboratory will not diminish. Increasingly sophisticated chemometric operations will be brought online with the instrumental source of the data, and those computations will be brought closer to taking place in real time.

Connectivity

Improvements in the technology and availability of networks will make a number of changes in the way computing develops. The wide-area networks are becoming faster, and this will make it practical, and eventually simple, to use a desktop computer as a front end to a database or to a supercomputer, possibly located across the country.

In one possible application, the results of large-scale computations performed on the supercomputer will be displayed on the desktop, which will be capable of very sophisticated graphics. Among other applications, remote databases might be queried, in some cases without the user needing to know where the data originate.

9.2 ALGORITHM DESIGN

What are the ramifications of the specific capabilities of desktop computers with respect to the code that is written? We will consider several factors: concurrency, parallelism and vectorizing, and bottlenecks.

Concurrency

A powerful desktop computer in the laboratory may have the capability to support both the instrument data acquisition and control tasks and the complex computational tasks, *if* appropriate priorities may be put into place.

In order to accomplish this, there must be a way of programming two tasks with a carefully controlled method of moving between the tasks. The data acquisition and control task will usually demand the highest priority from the computer because the acquisition of data with very high integrity is the most important task of the system. On the other hand, that task may not need to be monitored by the user, and therefore it may well be a background process.

Many desktop computers are not capable of directly supporting multiple concurrent tasks, largely because of limitations in the operating system software. Of those that are, most are not capable of *always* giving immediate, microsecond-level response to the background task. The best response might be achieved by a single-task computer that is programmed to service interrupt requests from the data acquisition subsystem. With such a scheme, data may be acquired with a time-base precision of a few microseconds while the majority of the computing cycles are given to data processing.

If parallel processing is possible, the CPUs might be assigned separate tasks. In one instructive example, a microprocessor machine was used to acquire data simultaneously with library searching [14]. However, this segmentation does not allow excess performance in one processor's task to be applied to the second task.

Parallelism and Vectorizing

Undoubtedly the potential for the highest level of numerical computing performance will come with the parallel and vectorized architectures that we may expect to see in desktop computers in the future. Efficient implementation in either case is not a trivial matter.

When parallel architecture is employed, the task of keeping all of the CPUs busy without resorting to manual optimization is challenging. Whereas adding the second processor may nearly double the throughput, the addition of more processors brings rapidly diminishing returns.

Vector architectures are much more successful than parallel architectures if the data vectors are large. The overhead of setting up a vectorized operation is significant, so that if the computer operates only on small vectors there may be a loss in throughput. The Cray examples in Table 1 illustrate the difference between the use of vector operations and nonvector (scalar) operations. If the processor were to perform only vector operations, the performance would probably be about three times as great.

Bottlenecks

Several operations on a desktop computer may present additional bottlenecks in the execution of data processing operations that can be solved by appropriate techniques.

Disk-intensive operations can dramatically slow down processing, even when very fast hard drives are employed. Cache memory for disk input/output can provide help by keeping in memory the most recently accessed areas of the disk; if the processing operation repeatedly goes to the same disk area, the likelihood of a "hit" is high, so no time is lost rereading sectors of information from the disk hardware. RAM drive (or RAM disk), an area of solid-state memory that appears to the operating system to be a very fast disk storage device, may provide useful temporary storage; the danger is that all of the information on a RAM drive becomes inaccessible when the system crashes or is lost when the computer's power is turned off.

Floating-point computations are the most common of the specialized applications that benefit from accelerators such as coprocessors. However, another technique that can be even more effective is the *look-up table*; if, for example, large numbers of sines are required *and* it is known that all of the angles for which the sines are to be computed are specified to 0.1°, an array of 900 real numbers, computed in advance, could contain all the results.

Graphics operations such as converting drawn vectors to rastered lines (i.e., rasterizing) and rendering or shading are time-consuming operations that can best be offloaded to special graphics processors.

9.3 SOFTWARE AND PROGRAMMING

It may seem extraordinarily obvious, but we must point out that a computer requires software, and the availability and quality of that software will frequently be the determining factor in the effectiveness of a small computer application.

General Categories of Software

Software for laboratory computers can be divided into three categories: the operating system, the application software, and the development tools.

First the computer requires an *operating system* (OS), which has three tasks.

THE ROLE OF THE MICROCOMPUTER

1. Communication with the specific I/O hardware of the computer, so that the user or programmer need not be concerned with such details as interacting with hardware to send characters to a screen or printer or to read and write information from and to a disk. It is possible for many computers with the same CPU to share software even though their I/O architectures may vary.
2. A set of utilities—a group of commands and/or programs that can handle the "housekeeping" associated with computer operation: copying disks, setting the clock, erasing files, and so on.
3. The monitor/executive, which handles the user's interaction and the scheduling of processes in multitasking systems.

Certain operating systems are specific to computer types: MS-DOS and OS/2 for computers using the 80x86 family of CPUs, operating systems for Apple II and for Macintosh computers, and CP/M for computers using Intel 8080 and Zilog Z80 CPUs. All are single-user, and most are single-tasking, operating systems.

On the other hand, Unix is a multitasking operating system that has been ported to a very wide variety of CPUs from small Intel 80x86 microcomputers to large IBM mainframe computers. Unix is extremely popular with professional programmers and has been the "operating system of the future" for about 20 years, providing a very rich set of utilities and powerful data handling features. However, to the less dedicated user, the commands appear cryptic, and Unix usually requires very large disk storage.

Unix was not originally designed for online applications, and slow response times are evident in most implementations, although real-time versions of Unix and real-time extensions are becoming increasingly available. The primary application for Unix has been in multitasking/multiuser computer systems without the need for real-time response. The largest number of Unix applications today are turnkey systems in vertical markets where a "user-friendly" shell or GUI conceals the operating system from the user. Unix is emerging as the operating system of choice in the growing workstation and "supermini" market and surely will continue to be a favorite among developers of computation-intensive applications.

The *application software* produces the end product of computing: the processed information. Generally speaking, there are two possible ways to obtain application software: obtain a program written by someone else, or write your own software. The former may not be possible, particularly when you are investigating new techniques. The latter is deceptively expensive, following the well-known 90-90 *rule*: "The first 90% of the task requires 90% of the time, and the last 10% of the task requires the other 90%." However, writing one's own software readily allows a user to validate and modify the code. Areas such as chemometrics, which are not yet mature, frequently require the users of the techniques to write their own code, even for well-understood methods.

Finally, *development tools* are required if one is to write software: an appropriate editor for entering the code, a compiler to convert that code to CPU instructions, a linker to combine different components of the code, and a debugger to find the errors in the operation of the program.

General Programming Languages

Possibly no subject relating to computers will elicit as much emotional response as the choice of the "best" programming language. In fact, there are several languages that have an important place in the scientific applications stable. In the following paragraphs, a few comments about some of these will be offered, but first a few general comments should be interjected. First, performance of the final code has as much or more to do with the quality of both the compiler and the application code as with the nature of the language; often there will not be significant differences in the execution speed of the same application written in different languages if quality compilers are employed. Second, among the considerations for the choice of language should be (1) the ease with which the programmer can express the concepts of the problem in a way that is easily understood and (2) the extent to which the structure of the language facilitates debugging and maintenance.

In the following paragraphs, a very brief summary of the features of common languages is presented. A more detailed discussion including code in each language for the same problem can be found in reference 1.

Assembly language presents two levels of uniqueness. First, it requires the programmer to explicitly code each and every discrete "machine-level" instruction executed by the CPU. Second, the language will be unique to the particular CPU. The primary advantages will usually be compact size and high speed because the programmer may recognize shortcuts that the compiler will not. However, it is seldom appropriate for chemometric operations because (1) coding is slow, simply because of the massive amount of code that must be generated; (2) debugging is difficult because so few errors can be detected by the assembler; and (3) the resultant code is not generally portable to other computers. On the order of 10 lines of assembler are required to replace an average line of FORTRAN or Pascal, and if the commonly accepted estimate is correct—that a programmer writes approximately the same number of lines of code per day, regardless of language—assembly language code development is very inefficient.

FORTRAN (*Formula Translator*) is the oldest of the computer languages still in scientific use. It retains wide popularity in the scientific community largely because of the extensive libraries of procedures that have been developed over several decades. Recently, many of the features of modern languages have been incorporated, thereby improving the efficiency of program development. However, were it not for its history and its built-in complex number capability, it would probably not be competitive today.

BASIC (Beginners' All-purpose Symbolic Code) is often the first language to become available for a microcomputer. In most current versions, the code can be compiled into an executable module; however, BASIC usually takes the form of an interpreter—when the program is run, each statement is compiled at the time it is executed. The consequence is great convenience at the expense of an order-of-magnitude slower execution. Until recently BASIC lacked many features that encourage well-organized programming. Some recent versions have overcome the older drawbacks, allowing named subroutines, data structures, and so on.

Pascal, a language named after the mathematician Blaise Pascal, was designed as a teaching tool and introduced rather recently, in 1968. Pascal is a "structured language," having features that are oriented toward code that is well organized and comparatively easy to maintain. The best known of these features is the near absence of the GOTO statement, which contributes to what is often termed "spaghetti code" in FORTRAN and BASIC. The organization also involves data structures, data blocks that can be manipulated as single variables. Recently, the concept of *object-oriented procedures* of OOPS, has become popular, allowing larger processing modules to be manipulated as a single structure. Pascal is limited in the number of built-in mathematical functions. Possibly its greatest limitation is that Pascal lacks some standardization, particularly in handling input and output.

Currently, Pascal is the most popular language for teaching programming on university campuses because, as a structured language, it forces good programming practices. Several microcomputer implementations [e.g., TurboPascal (Borland International, 4585 Scotts Valley Dr., Scotts Valley, CA 95066)] provide a single editing, execution, and debugging environment while producing code that executes very rapidly.

C is a compiled language whose development was related to the development of Unix at Bell Laboratories. Its capabilities bring the programmer much closer to the CPU instruction level than do the languages mentioned so far. It has a rich set of direct and indirect addressing techniques and a full set of instructions for bit manipulations. Consequently, C is typically the language of choice for writing operating systems, other compilers, word processors, communications software, and so on—software that must operate efficiently and is hardware-oriented.

The high degree of flexibility also leads to more opportunities for errors in C, and consequently the debugging support (see the section on Debugging) is developed to a higher level than is true for other languages.

C programming has many of the benefits of assembly language programming, but the code is extremely portable; C compilers that conform to accepted standards are available for most CPUs that are used in scientific applications. On the other hand, several disadvantages must be borne in mind: Mathematical capabilities are limited in most implementations, the terse syntax can be difficult to master, and debugging can be more difficult than for Pascal.

FORTH was developed primarily as an instrument control language and has a history and style quite distinct from other recent languages. FORTH is based on the concept of a dictionary; entries (or *definitions*) are reference procedures that are written using other entries and effectively become part of the language. Parameters are passed between these executable modules via a last-in-first-out stack. As a consequence, single definitions can be executed interactively or strung together until the entire program is a single definition.

FORTH has been very popular for instrumentation applications because it is efficient in its use of memory, single definitions can be developed and tested very rapidly, and its stack-oriented approach makes procedures inherently reentrant; consequently, context switching to handle real-time events is fast and easy [5]. However, its popularity appears to be in decline because management of the dictionary and maintenance of the code are very difficult. The syntax, including reverse Polish notation for arithmetic procedures, is radically different from that of other languages.

One implementation of FORTH, called ASYST (ASYST Software Technologies, Inc., 100 Corporate Woods, Rochester, NY 14623), does include an extremely rich array of functions for graphics, matrix algebra, and statistical functions [15]. Consequently, it has developed a significant following among scientific users.

Specialized Languages

Special-purpose languages have been developed for tasks such as mathematical manipulation, finding numerical solutions, algebraic manipulation, and real-time interfacing. Since the "code" frequently does not look like that of conventional languages, many vendors are unwilling to call them languages, even though they are high-level methods for encoding algorithms. Some of these tools may be of relevance to this discussion.

Some very powerful tools for *interactively* manipulating data have changed the way many users program. MATLAB (The Math Works, Inc., 21 Elior St., South Natick, MA 01760) and MathCAD (MathSoft, Inc., One Kendall

Square, Cambridge, MA 02139) are two approaches to providing an interactive tool for the manipulation of data with over 100 functions built in. C code can also be linked into MATLAB to extend the tool for computational or data acquisition purposes.

In an effort to make interfacing of laboratory peripherals accessible to those who would not consider themselves to be programmers, an "icon-based" method has been developed by several vendors [16]. In each case, an icon representing an interfacing tool (analog-to-digital converter, clock, panel meter, binary input, disk file, etc.) is dragged to the screen by a mouse. These icons are then connected by drawing lines to indicate function. Although complex real-time control may be very difficult to program by this method, simple acquisition functions can be developed in minutes.

Debugging

"Bugs," or errors, in computer operation have been part of the terminology ever since a moth was found to be the fault of errors in an early computer that employed mechanical relays as memory. Until recently, the problem of finding the errors in a program received almost no formal attention in spite of the acknowledgment that half of programming time is spent in testing and debugging. A recent treatise on the subject addresses the problem both in general and for the particularly difficult problem of debugging in C [17].

Although compilers may catch most (but not all!) lexical and syntactic errors, there are other errors that may show up only during execution and then often only after aggressive testing. (As programmers, we may find ourselves subconsciously saying "nobody would want to do that" instead of dispassionately testing all combinations of input.) Errors of intent may be even more insidious: The code may do exactly as the programmer wanted, but the programmer wanted the wrong thing.

For these bugs, a scientific, systematic approach is required [1, 17]. While the development of the code may allow a programmer to operate as a skilled craftsperson, debugging requires a skilled scientist, considering that "[the] scientific method seeks to understand the unknown, to explain observable natural phenomena, and to expose cause-and-effect relationships through repeatable experiments . . . [and] bugs are unknown (initially) natural phenomena whose cause we want to understand" [17].

There are several tools available to the programmer to be used as instruments to study the code. A logic analyzer is a hardware tool that will trace the actual cycles of the computer and is useful if the source code at the assembly level is available *and* the hardware is understood. Typical debugger software can also be useful, again if the source code at the assembler level is available, to allow one to stop execution at any point in the code and examine the status;

unfortunately, it loses some of its value when the code is being debugged for real-time operation. Recently, "symbolic" source code debuggers have become available to operate with the compiler so that the advantages of debuggers can be obtained without the requirement of assembly-level code.

In the end, however, the most important component is a systematic approach. This begins with careful and "clean" code design, including rigorous error-trapping where appropriate, followed by debugging. Each debugging step should be designed as an experiment—a probe of the software whose results will yield some contributing information.

9.4 LABORATORY INTERFACES

In the past, intensive chemometric operations were performed on offline computers separate from data acquisition. However, the voluminous data and a need for real-time or near-real-time processing requires that the laboratory computer perform both operations. A study of the interface between the computer and the experiment can be made from that perspective.

In each case, the information moves from the physical/chemical domain to digital information through processes of interdomain conversions [18], illustrated in Figure 6. First, a *transducer* converts the physical/chemical phenomena to the electrical domain—usually voltage, current, charge, or resistance; transducers exist for the measurement of nearly every phenomenon [1]. Second, that electrical information will require *analog processing*, typically amplification but possibly the application of correlation methods, rectification, or conversion to other electrical domains. Third, the information undergoes *analog-to-digital* (A/D) *conversion*, placing the data into the digital domain. Finally, the *data acquisition* operation places it into computer memory.

Analog Interfaces

Most experimental information is initially nonquantized, at least in easily measurable quantities such as temperature, light intensity, and electrical potential or current. The transducers in most cases generate analog information in the elec-

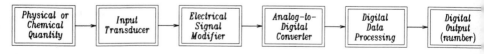

Figure 6 Schematic illustrating the domain conversions of basic digital electronic measurement. (Adapted with permission from reference 18, copyright © 1971, the American Chemical Society.)

THE ROLE OF THE MICROCOMPUTER

trical domain. With suitable analog processing, that information may be encoded as a voltage in the 0–10-V range, suitable for A/D conversion.

The designer of the interface chooses from several types of analog-to-digital converters, determining which best fits the needs of the signal. The methods of conversion are best studied elsewhere [1], but the characteristics of each are important.

The *successive-approximation* (SA) ADC is the most versatile and most common. A successive-approximation process, requiring approximately 1–1000 μs, leads to a digital result with a typical resolution of 12 bits (1 part in 4096). The conversion process begins with a sampling operation, acquiring the analog signal for subsequent conversion by sampling over a period of tenths to tens of microseconds.

The *dual-slope ADC* and the *voltage-to-frequency ADC* both integrate the signal over at least part of the conversion period. Typically, the conversion period is relatively long, 6–33 ms. Resolution of 12 bits is again typical, but 16-bit ADCs usually employ dual-slope architecture.

Flash ADCs occupy the other end of the spectrum. Used most often for acquisition of video information, they can convert at rates up to 500 MHz. Flash ADC resolution seldom exceeds 8 bits because of the huge scale of semiconductor integration that is required. Furthermore, no computer could keep up with those rates if the data had to pass through the CPU, and specialized local memory is required.

The SA ADC is most frequently encountered in general-purpose analog interfaces, where it is part of a subsystem of components as outlined in Figure 7. The SA ADC must be joined by another component, a *sample-and-hold amplifier*, which is required to guarantee that the voltage is stable during the conversion period. Because the ADC is often large and expensive, it is usually shared by several analog signals, requiring a *multiplexer* as a switch. Optionally, a *programmable-gain amplifier* can be added; the computer controls the gain of an amplifier to ensure that the full range of the signal fills the capacity of the ADC.

When a signal is sampled periodically, one runs the risk of *aliasing* the signal, that is, acquiring the signal at too low a frequency so that the result has the appearance of a much lower frequency; to avoid aliasing, the data must be acquired at a frequency greater than twice that of the highest frequency com-

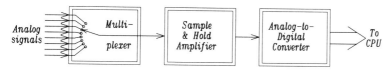

Figure 7 Components of a successive approximation ADC subsystem.

ponent of the signal (the Nyquist frequency). The example in Figure 8 illustrates the result of acquiring a sine wave at rates below the Nyquist frequency.

The greatest threat from aliasing arises when the acquisition rate exceeds the Nyquist frequency of the true signal but systematic noise at higher frequencies corrupts that signal. The noise can be aliased down to frequencies comparable to that of the signal, and that noise can be very resistant to removal by computer-based digital filtering techniques. The example in Figure 9 illustrates an exponential decay corrupted by high-frequency noise in which the noise is aliased to a low frequency that greatly distorts the signal.

Standard Interfaces

There are a few agreed-upon standards for interfaces. These may be de facto standards or industry standards. A *de facto standard* is one in which the interfacing technique used in one product is followed by the remainder of the industry. Examples include the parallel printer interface developed by Centronics; it is now followed by most printer manufacturers and is known simply as a Centronics interface. When a printer is specified to have a parallel printer interface, it can be connected reliably with a microcomputer. Another example of a de facto standard is the bus specification for the IBM PC computer.

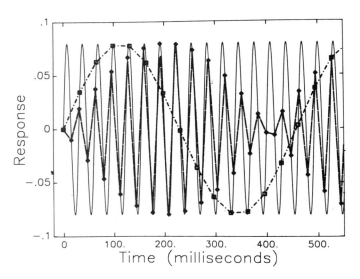

Figure 8 Aliased acquisition of a sine wave. The solid line is the actual signal, and the broken lines are aliased representations. (Reprinted from reference 1 by permission of John Wiley & Sons, copyright © 1987.)

THE ROLE OF THE MICROCOMPUTER   295

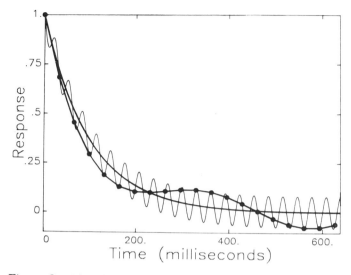

Figure 9 Aliased acquisition of an exponential decay curve corrupted by sinusoidal noise. (Reprinted from reference 1 by permission of John Wiley & Sons, copyright © 1987.)

An *industry standard* is one that was carefully defined by an independent body, most commonly the IEEE (Institute for Electronics and Electrical Engineers). In many cases, an IEEE standard grows out of a de facto standard as many vendors of electronic equipment seek to ensure reliable connectivity between their instruments and computers. There are many standards for buses and input/output modules. The two that are most common for interfacing to microcomputers follow in this section.

General-Purpose Instrument Bus (GPIB)

The GPIB standard, defined by IEEE-488, is an industry standard that grew out of the Hewlett-Packard interface bus. It is a true standard in that it has been defined with great precision so that a wide range of instruments from many manufacturers can be expected to connect together both physically and electrically and to operate immediately.

The specifics of IEEE-488 are quite detailed and complex, beyond the scope of a simple discussion [19, 20]. However, interface boards are readily available and include software drivers that make application quite straightforward. The IEEE-488 interface is a bus in that a single set of conductors share signals from a large number (up to 15) of devices. The data bus is 1 byte wide for parallel transfer. Each device has its own address but uses the same 24-

conductor stackable connector, outlined in Figure 10. The address is usually specified on each device by a set of switches.

Devices that share the IEEE-488 bus must be one of four types: talker, listener, talker-and-listener, and controller. A *listener* can only accept data; examples might be a digitally controlled power supply or waveform generator. A *talker* will only send data; a digital voltmeter and a pH meter are examples. A *talker-and-listener* can do both, and most laboratory instruments fall into this category; a typical spectrophotometer will both generate data (talk) and accept operational parameters (listen). Finally, a *controller* can not only talk and listen but also send commands to other devices; the controller may control other instruments as well as talk and listen. Although a controller may be absent from the IEEE-488 bus (for example, when a recorder logs data from a talker), there may not be more than one active controller on the bus at a time. The small computer usually assumes this role.

The advantages of this method of interfacing are that (1) it is well defined; (2) many instruments can share the same interface port on the computer; (3) it is relatively fast, since the data lines are organized in parallel; and (4) it is increasingly common. On the other hand, it is rather complex in hardware and

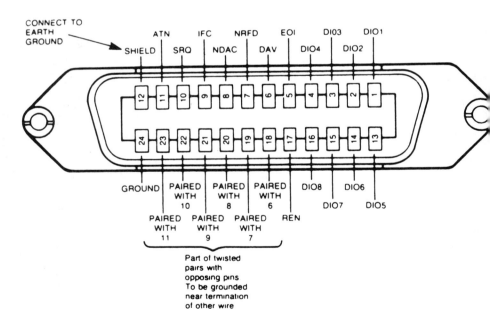

Figure 10 Connector used with devices following the IEEE-488 standard. (Reprinted from reference 1 by permission of John Wiley & Sons, copyright © 1987.)

THE ROLE OF THE MICROCOMPUTER

software (although that should be hidden by well-designed drivers) and is moderately expensive compared with serial and printer ports.

RS-232 Asynchronous Serial Communications

This standard [6] is the most common for connection between instrument and computer, but unfortunately it is far from being a true standard; it is seldom possible to connect cables that are physically compatible and expect that communications will operate. A brief look at the standard and how it is used may be helpful, but a detailed look at the specification of the standard is not.

The term *serial* communications refers to the timing of the transmission of the bits in a single byte: the bits are transmitted sequentially on a single conductor (as compared with a parallel interface, which requires one conductor per bit). Serial communications are common because data can be transmitted bidirectionally with only two conductors (and ground), or over the phone line, or by fiber-optic lines. In most microcomputers, the communications are also *asynchronous*, meaning that there is no synchronization between the communicating devices to determine when a new datum will be sent. The primary disadvantage is the slow speed of transmission, necessitated by synchronization and serialization of the data.

Between the devices at the two ends of the RS-232 serial communications cable, there must be agreement on a large number of specifications, the most important of which are summarized in Table 2. There are many other pins with specifications for specialized applications [1, 6], but these are rarely used in microcomputer applications.

Optionally, a parity bit can be added to the datum so that the sum of all the bits will be either even or odd. This bit, which was mandatory when transmission over noisy phone lines was common, can detect (but not correct) a single bit received in error.

Handshaking (flow control) is necessary if the receiving device (e.g., a plotter) cannot always handle the data as fast as they are received. Separate lines can be used, such as the DTR/DSR pair, to signal when the device is ready, or the receiving device may send transmit control characters back to the sender; characters denoted X-On and X-Off are the most common.

The person setting up the communications must ensure that the two devices are matched, particularly to determine the following:

Which device is DTE and which is DCE (in order to determine pin numbers)?
What is the baud rate?
How many bits per datum are to be transmitted?
What, if any, parity bit will be added to the end of datum to help check for errors?
What kind of handshaking is to be used?
What is the minimum time (measured as "stop bits") between transmitted bytes?

Table 2 Commonly Followed RS-232 Specifications

Which bit goes first	Least significant
Number of bits per second (baud)	Usually 110, 300, 600, 1200, 2400, 4800, 9600, 19.2K, or 38.4K
Logical level at idle	High
Number of bits transmitted per datum	Usually 7 or 8
Type of connector	D-subminiature, 25-pin; female for DCE; male for DTE; 9-pin becoming common.
Pin on which data are sent	Pin 2 for DTE; pin 3 for DCE
Pin on which data are received	Pin 3 for DTE; pin 2 for DCE
Voltage levels to represent a '1'	-5 to -25 volts
Voltage levels to represent a '0'	$+5$ to $+25$ volts
Handshake line to signal "ready"	Pin 20 (DTR) for DTE; pin 6 (DSR) for DCE. (Pins 4 and 5 are occasionally used in a similar manner.)
Handshake line to monitor "ready"	Pin 6 (DSR) for DTE; pin 20 (DTR) for DCE. (Pins 4 and 5 are occasionally used in a similar manner.)
Method of signalling the beginning of a byte	A logical '0' for the period of one bit
Error detection	None, even parity, or odd parity

DCE = data communications equipment; DTE = data terminal equipment; DTR = data terminal ready; DSR = data set ready.

Typical violations of the standard by instrument manufacturers include the failure to use the sex of connector that corresponds to the choice of DCE or DTE pin designations and failure to use pins 6 and 20 as a handshaking pair (mixing them with the pair composed of pins 4 and 5).

After being very careful to set baud rate, word length, and parity to be the same at both ends, and determining that all active sending pins on one device are routed to receiving pins on the other, the user will usually meet with success, at least with the transmission of the raw characters.

Parallel Digital Interfaces

The foundation of any interface is the component that is the gate between the digital information and the bus of the computer. There are two occasions on which a user might need to work with such an interface. First, when the data are presented by an instrument in a fully parallel fashion; for example, a 4½-digit pH meter might have a pin for each of the 17 binary lines required to encode the

THE ROLE OF THE MICROCOMPUTER

pH. Instrument houses no longer take this approach, although it was common a decade ago. Those 17 lines may be multiplexed into a single 8-bit port so that the computer selects each 8-bit group and then reads that part of the datum separately.

When parallel data are transferred, one must pay careful attention to handshaking, as suggested in Figure 11. The receiving device may need to indicate when it is ready to receive a datum, and the sending device, upon determining that the receiver is ready, must indicate that a new datum is being sent by generating a strobe pulse.

The other type of occasion that may call for a parallel digital interface is when an experiment requires a collection of bits for control purposes: for example, bits to control relays or motors, or to sense proximity. Many of these single bits will need to be manipulated one at a time without handshaking.

Timing in Real-Time Operations

A real-time system must operate within two timing constraints. First, it must have a highly precise measure of the time, a task typically handled by a hardware counter and a crystal-based oscillator. Measurements of the magnitudes of signals in events in many scientific experiments are usually performed with precision, but frequently insufficient attention is paid to the time base. Second, when the computer responds to an event (a tick of the hardware counter or a flag from an external device), it must do so with a predictable and short delay, even in the worst case [5].

The methods for input from and output to external devices fall into three categories: polling, interrupts, and direct memory access. The methods

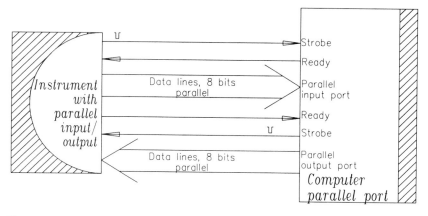

Figure 11 A digital interface showing the handshaking lines.

can be examined from the aspects of speed, efficiency, and hardware/software complexity.

When *polling* is employed, a loop must be placed in the program to execute during the time that a response may be needed. As illustrated in Figure 12, the loop continuously examines the device to determine when it is ready. For example, when a clock pulse is used to trigger an ADC, the computer must

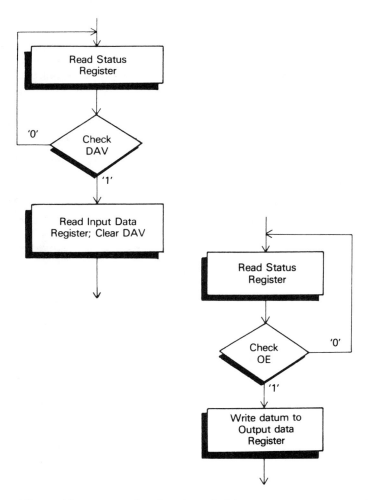

Figure 12 Program flowchart for polled data acquisition. (Reprinted from reference 1 by permission of John Wiley & Sons, copyright © 1987.)

THE ROLE OF THE MICROCOMPUTER

respond to the end-of-conversion (EOC) signal by reading the results. A loop must be employed to poll the EOC signal, test it, and read the datum when the EOC is TRUE.

When *interrupts* are used, the software procedure for responding to the external device is called by the hardware. An interrupt request is made in the form of a pulse routed to the CPU, and at the end of the current machine instruction, control is immediately transferred to the *interrupt service routine*, which can take immediate action. At the end of that procedure, as illustrated in Figure 13, the main routine resumes operation where it left off, oblivious to the interrupt activity. Some overhead is required before the interrupt service routine can operate; the context of the main routine (the contents of the CPU's internal registers) must be saved in memory to be re-stored at the end of the routine. Returning to the ADC example, the EOC flag can be used to request an interrupt, and the interrupt service routine will acquire the datum, store it in an array, and set a flag to indicate to the main routine that new data have been

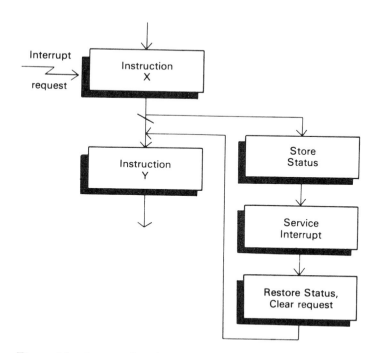

Figure 13 Program flowchart for interrupt-driven data acquisition. (Reprinted from reference 1 by permission of John Wiley & Sons, copyright © 1987.)

acquired; whenever convenient, the main routine can examine that software flag and take the indicated action.

For most current CPUs and operating systems, the time required for context switching following an interrupt service request is not fixed [5], causing substantial variability in the response time. When an event requires attention, the worst-case response time, not the average, is important. The variability comes both from the CPU, the compiler, and from the operating system. The number of cycles required by the CPU for completion of the current instruction in the main program on a CISC CPU can vary from one or two to over 50. The compiler must determine how many registers to save before the service routine can be executed (one CPU has 192 internal registers, not all of which might be affected). Furthermore, at times that are difficult to control, the operating system may turn off all interrupt activity of a given set of tasks in order to handle other interrupts or time-dependent tasks. Only when other activities (principally user, disk I/O, and time-of-day clocks) are turned off can the response time become reliable.

Direct memory access (DMA) is appropriate when data are to be input or output with no intelligent action required [21]. The DMA controller bypasses the CPU altogether and transfers a datum directly to or from the correct memory location. The transfer takes place either at the next opportunity when the bus is unused by the CPU (cycle stealing) or by placing the CPU temporarily on hold before the burst of activity.

These methods can be compared using the ADC input as an example. From the aspect of throughput, DMA will usually be the winner; the clock cycles required to move the datum into the CPU, for the CPU to determine the destination, and for the datum to be moved there are eliminated and replaced by a single-cycle transfer coordinated by a DMA controller. Depending on the architecture of the computer (particularly that of the memory), the upper rate for DMA will be 100 kHz to several MHz (often limited by the need to refresh dynamic memory). Polled I/O may proceed at 20–50 kHz, while the overhead of context-switching slows interrupt procedures to 5–20 kHz. (In each case, it is assumed that no processing of the data takes place in real time except to transfer from port to memory.)

Efficiency is another matter. Under both DMA and interrupts, the main routine need not be coordinated with the timing of the input/output, allowing generation of much more efficient code. Data may be processed in near real time without the need to guarantee that data are processed as fast as they are received. The exception occurs when a real-time response to an input datum is required; that action is not ensured under DMA but may be programmed into an interrupt service routine.

Finally, there is a substantial difference in the complexity of hardware and software. DMA requires specialized hardware. However, in an IBM PC/XT or

PC/AT, uncommitted DMA channels are built into the system; only small changes to the I/O board are required. Nevertheless, DMA remains an expensive option for commercial ADC boards. Software for setting up the channel is also required; currently, operating systems and compilers do not explicitly support user DMA, and a working knowledge of the DMA controller [21, 22] is mandatory. The hardware for interrupts is a part of most input/output systems, and increasingly compilers implement explicit support; an important caveat, however, is to disable interrupts after use so that an inadvertent interrupt service request will not crash the computer after the program is terminated.

In summary, interrupts offer the most flexibility because they are efficient and a substantial amount of processing can be placed in the service routine while the main software system operates asynchronously. Polling *can* be faster, but for moderately slow data the technique is very inefficient. DMA is the fastest and most efficient, but only when data are to be transferred without processing in real time.

9.5 LABORATORY INFORMATION MANAGEMENT SYSTEMS

A system for managing the information generated in a laboratory is generally categorized as a *Laboratory Information Management System* (LIMS). An exact definition may be difficult to defend, but in a discussion by Megargle [23], a LIMS is described as an "encompassing system that includes some or all of the functions of test ordering, sample tracking, work-station management, results reporting, archiving, quality assurance, and the gathering and reporting of laboratory management data."

A LIMS will generally be associated with a laboratory that performs a high volume of chemical or physical tests that can be defined in advance even though the correlation and interpretation of the results may be more complex. Examples of laboratories that make effective use of a LIMS include quality control laboratories, research laboratories with large numbers of routine tests (e.g., pharmaceutical laboratories), and regulated laboratories.

One aid to understanding the function of a LIMS might be the typical installation shown in Figure 14. A centralized computer interacts with multiple terminals, instruments, and a computer network to gather information and make it available to users at the terminals and to the printing and storage devices.

A function that is common to every LIMS is that of gathering and manipulating laboratory data. In many systems, the data are entered by hand at a terminal by the analyst, but increasingly intelligent instrumentation enables the LIMS to obtain that information directly from the instrument or at least from a small computer interfaced to the instrument.

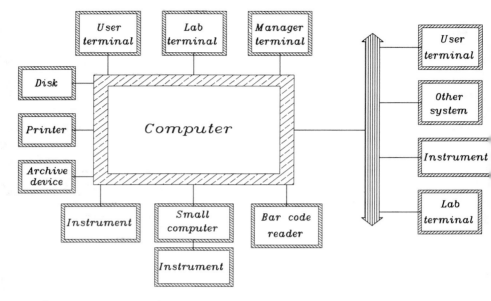

Figure 14 Typical hardware for a laboratory information management system. (Adapted with permission from reference 23, copyright © 1989, the American Chemical Society.)

From the data, the LIMS can generate reports or control charts for single samples that have been subjected to multiple tests, or for large volumes of samples, depending on the need of the user.

Many laboratories benefit from automated sample entry and tracking, often including bar code readers at each experimental workstation. By freeing the analyst from numerical entry of either test results or sample numbers, the opportunity for error is greatly reduced. At the same time, the computer can be used to order the tests to be performed and inform the analyst at each workstation of the samples to be run.

Possibly the most common form of LIMS is the spreadsheet, a tool developed originally for the business community as an "electronic accounting book" but finding increasing application in the sciences [24]. Data and other information may be placed in a grid, and operators can be defined for operations on particular cells in the grid. Modern spreadsheets have powerful capabilities for graphics, data analysis, and report generation [25]. Using clever techniques and/or add-on capabilities, the movement of data from instruments to the spreadsheet can be substantially automated. However, when the number of data sources exceeds a half dozen and the number of locations needing access to the data exceeds one, the spreadsheet will lack the needed flexibility. A full-scale LIMS will be required.

For an analytical laboratory to be efficient and productive, it is necessary to use a LIMS, but it must be apparent that the installation of a LIMS can be quite complex because it must communicate with such a diverse array of devices and produce many different kinds of output. The need for standardization is being addressed by an ASTM committee [23]. The interaction of local area networks will also be necessary for the effective movement of information.

Finally, the interpretation of results, particularly when the result of different tests must be combined to make a decision, will bring chemometric methods (for correlating multisensor information about the same problem) and artificial intelligence to the future LIMS. As robotics have matured, we can expect that the results of the decision making in the software will be carried out by this programmable hardware.

9.6 ARTIFICIAL INTELLIGENCE ACTIVITIES WITH MICROCOMPUTERS

It has long been the desire of scientists to "endow computer-controlled machinery with the ability for actions which, if done by a human being, would be thought to require intelligence" [26, 27]; that general characteristic is known as *artificial intelligence*. Within that large field lies a subdiscipline, *expert systems*, that has achieved a significant level of success. Another area, *neural networks*, has captured the imagination of a number of chemists and has produced some successful applications. Many other areas of artificial intelligence show varying degrees of promise but are outside the scope of this discussion.

Expert Systems

An expert system [28–31] is a computer program that can perform some of the problem-solving tasks of a human expert. The solution is a two-step process. The first is *knowledge engineering*, which is the task of converting the knowledge embodied in an expert to formalized rules [32]. For a particular subject under study, the rules are gathered from an expert and codified by a "knowledge engineer." The second step is to be able to draw conclusions or inferences from the combination of those rules and a set of facts; this requires a program called an *inference engine*. A third component that is assumed essential is a *human–machine interface* [33].

In their simplest form, the rules are a series of IF-THEN-ELSE statements that can be combined (chained) using Boolean logic to make an inference chain [28]. Frequently, however, the IF-THEN-ELSE may include a "probably," in which case *fuzzy logic* may be involved to deal with probabilities.

When small numbers of rules are required, simple programs in any common computer language can be used to hard-code those rules and apply them to the data at hand. However, as the number of rules gets larger, the rules should also be treated as input information rather than program code. The construction of the inference engine can benefit from languages adapted for that purpose: LISP and PROLOG are the major examples. The body of rules may be dynamic as rules are changed or improved, but the inference engine can readily operate on the current rules, which, together with a set of facts, leads to appropriate new conclusions.

The application of expert systems to chemistry has covered areas in which expertise is in demand but an insufficient supply of experts exists: examples include high-performance liquid chromatography [34], mass spectrometry [26], and ultracentrifugation [35]. When one views robotics as programmable hardware analogues to the computer as programmable logic, the importance of current efforts in that area becomes apparent [36].

Neural Networks

The concept of a neural network [37, 38] is loosely based on a simplified model of the brain in which a number of processing nodes are connected in a network structure. One such network is outlined in Figure 15.

Information is fed into the input layer nodes encoded as signal strengths—analog or digital, depending on implementation and purpose. Signals are issued from one or more nodes in the input layer indicating "decisions" or result(s); one or more "hidden" layers may be used in the processing path. Information processing occurs as each node is fed from nodes of earlier layers with values that are weighted and summed. This weighting of information is the processing function or "program" that assigns the significance the input information has on the output result(s).

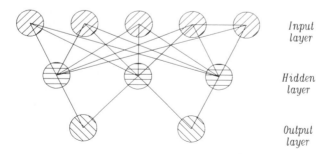

Figure 15 Schematic representation of a neural network. (Adapted with permission from reference 37, copyright © 1989, the American Chemical Society.)

A key feature of the neural network is the capacity for self-learning without explicit entry of rules, unlike expert systems. In a learning process resembling an iterative minimization method, "training sets" of data are entered while internode feedback mechanisms adjust the weights to meet the known or desired results, until the weights become stable. The weights may then be fixed, or in some cases the network may continue to learn as it processes new data.

Although most work is currently performed by simulating the nodes on a single processor, neural networks may provide the potentially important benefit of profiting from parallel processing since each node acts independently. Workers with microcomputers may add neural-network coprocessor boards to the system to achieve these results.

REFERENCES

1. K. L. Ratzlaff, *Introduction to Computer-Assisted Experimentation*, Wiley-Interscience, New York. (1987).
2. P. Freiberger and M. Swaine, *Fire in the Valley*, Osborne/McGraw-Hill, Berkeley, CA. (1984).
3. K. L. Ratzlaff, *Am. Lab.*, 10(2): 17–27 (1978).
4. S. P. Perone and D. O. Jones, *Digital Computers in Scientific Instrumentation*, McGraw-Hill, New York. (1973).
5. R. Wilson, *Comput. Design*, 28(3): 88–105 (1989).
6. E. A. Nichols, J. C. Nichols, and K. R. Musson, *Data Communications for Microcomputers*, McGraw-Hill, New York. (1982).
7. R. L. Deming and L. A. Young, *Sci. Comput. Automat.*, 5(4): 21 (1989).
8. C. E. Catlett, *Acad. Comput.*, 3(5): 18–21 (1989).
9. S. R. Heller, *JCICS*, 29: 135 (1989).
10. S. R. Heller, *JCICS*, 29: 135–136 (1989).
11. A. G. W. Cameron, *MIPS*, April 1989, p. 18.
12. R. March, *PC Week*, 6(25): 77–79 (1989).
13. F. Hayes, *Byte*, 14(5): 113–114 (1989).
14. R. E. Fields III and R. L. White, *Anal. Chem.*, 59: 2709–2716 (1987).
15. K. L. Ratzlaff, *JCICS*, 29: 128–129 (1989).
16. *Labtech Notebook* from Laboratory Technologies, 400 Research Dr., Wilmington, MA 01887, *The Analog Connection Workbench* from Strawberry Tree, 160 S. Wolfe Rd, Sunnyvale, CA 94086 and *SciBench* from IBM, 44 S. Broadway, White Plains, NY 10604.
17. R. Ward, *A Programmer's Introduction to Debugging C*, R&D Publications, Lawrence, Kan. (1989).
18. C. G. Enke, *Anal. Chem.*, 43(1): 69A (1971).
19. Hewlett-Packard, *Tutorial Description of the Hewlett-Packard Interface Bus*, Hewlett-Packard, Palo Alto, CA (1980).
20. S. Leibson, *Byte*, 7(4): 186–208 (1982).

21. P. D. Wentzell, S. J. Vanslyke, and A. P. Wade, *Trends Anal. Chem.*, 9(1): 3–9 (1990).
22. L. C. Eggebrecht, *Interfacing to the IBM Personal Computer*, Howard W. Sams, Indianapolis. (1983).
23. R. Megargle, *Anal. Chem.*, 61: 612A–621A (1989).
24. M. L. Salit, *Anal. Chem.*, 60: 731A–741A (1988).
25. G. I. Ouchi, ed., *Laboratory PC User*, BREGO Research, San Jose, CA 95124 (1986).
26. C. G. Enke, A. P. Wade, P. T. Palmer, and K. J. Hart, *Anal. Chem.*, 59: 1363A–1371A (1987).
27. N. Graham, *Artificial Intelligence—Making Machines Think*, Tab Books, Blue Ridge Summit, PA (1979).
28. A. R. de Monchy, A. R. Forster, J. R. Arretteig, L. Le, and S. N. Deming, *Anal. Chem.*, 60: 1355A–1361A (1988).
29. T. H. Pierce and B. A. Hohne, eds., *Artificial Intelligence Applications in Chemistry*, American Chemical Society, Washington, D.C. (1986).
30. W. L. Truett, *Am. Lab.*, 21(4): 40–47 (1989).
31. F. A. Settle, Jr., and M. A. Pleva, *Am. Lab.*, 20(10): 64–79 (1988).
32. C. E. Riese and J. D. Stuart, A Knowledge-Engineering Facility for Building Scientific Expert Systems, in *Artificial Intelligence Applications in Chemistry* (T. H. Pierce and B. A. Hohne, eds.), American Chemical Society, Washington, D.C., pp. 18–36. (1986).
33. D. H. Smith, Artificial Intelligence: The Technology of Expert Systems, in *Artificial Intelligence Applications in Chemistry* (T. H. Pierce and B. A. Hohne, eds.), American Chemical Society, Washington, D.C., pp. 1–17. (1986).
34. R. Bach, J. Karnicky, and S. Abbott, An Expert System for HPLC Methods Development, in *Artificial Intelligence Applications in Chemistry* (T. H. Pierce and B. A. Hohne, eds.), American Chemical Society, Washington, D.C., pp. 278–296. (1986).
35. P. R. Martz, M. Heffron, and O. M. Griffith, An Expert System for Optimizing Ultracentrifugation Runs, in *Artificial Intelligence Applications in Chemistry* (T. H. Pierce and B. A. Hohne, eds.), American Chemical Society, Washington, D.C., pp 18–36 pp. 297–311. (1986).
36. T. L. Isenhour, S. E. Eckert, and J. C. Marshall, *Anal. Chem.*, 61: 805A–814A (1989).
37. S. Borman, *C&EN*, 67(17): 24–28 (1989).
38. D. S. Touretzky and D. A. Pomerleau, *Byte*, 14(8):227–233 (1989).

10
Trends in Chemometrics

Steven D. Brown   *University of Delaware, Newark, Delaware*

The preceding nine chapters have provided some insight into the philosophy and the mechanical aspects of chemometrics. The methods discussed in this book have increasingly been used to improve analytical measurements, to assist in evaluating the quality of data, and to permit quantitative estimates of concentrations from the overlapped responses of mixtures. Many of the chemometric methods presented in this book have been known for some time, but they were not used until recently. Improved access to cheap, fast computers has made feasible many mathematical and statistical methods that were previously thought to be computationally inconvenient for use in the analytical laboratory. It is now as simple to perform principal component analysis or to calibrate a multicomponent system as to calculate a standard deviation. This power must change the way chemists look at data. As the previous chapters have shown, a multivariate approach to data analysis offers access to information that until quite recently was invisible to the casual observer.

These chapters have also provided the reader with some feeling as to the direction of research on new chemometric methods. New measurement techniques provide large amounts of raw data. Evaluating that data and extracting the information locked within it is a task that is ever changing as the sophistication and capabilities of modern instrumentation increase. In some cases, statistical and mathematical methods have not been developed for handling the more complex (and usually larger) data sets available from these instruments.

These factors drive research into new methods of chemometrics. The goal of this research remains the same as with the application of older methods: the efficient extraction of information from the data.

10.1 CHEMOMETRICS AND INFORMATION THEORY

All of chemometrics, application and method development, concerns the conversion of large amounts of chemical data into somewhat smaller amounts of useful results—the "information content" of that data. To this point, the concept of information content has only been considered loosely, as in the preceding two paragraphs. We can be much more specific about this conversion, however. It is useful to regard this conversion process as a kind of communication, from the experiment to the analyst. The data serve to transmit the "signal" containing the quantities sought by the analyst in performing the experiment. Using a theory developed for evaluating communications, we can consider the efficiency of extraction of information more quantitatively. Shannon [1] demonstrated that the gain in information produced by a transmission can be related to the reduction of uncertainty, where the uncertainty is closely related to classical probability theory.

Information Theory: A Brief Introduction

We can gain some appreciation for Shannon's theory by considering how information theory is used in two simple problems: species identification and species quantitation [2]. Consider the problem of identifying the single cation present in a sample that could contain any one of n cations through the classical qualitative analysis scheme. We begin the sequence of tests with no other prior knowledge of the sample. Now, from classical probability theory, each ion x_j in the scheme has a chance $p(x_j)$ of being present. The uncertainty measured before the qualitative analysis scheme is reflective of the fact that each ion is either present or absent, and each choice has some probability. Shannon defined this initial uncertainty in terms of the entropy, H, where

$$H_{\text{before}} = \sum_{j=1}^{n} p(x_j) \log_2 [p(x_j)] \tag{1}$$

For the moment, we will assume an unrealistic situation: each of the n cations for which we will test has an equal chance of being present. We make this assumption because we have no prior "information" about this sample that might alter our assessment of probabilities. When we add HCl and observe the result, we learn of the presence of Ag^+, Hg_2^{2+}, and Pb^{2+} ions. A precipitate communicates to the analyst that one (or more, in general) is present. The absence of a precipitate communicates that none of these three ions is present but provides no help on the presence or absence of other ions, say Cd^{2+}. The uncertainty after the experiment must reflect these facts in assigning probabilities, based on the results of the experiment. Thus, the new uncertainty is

$$H_{after}(k) = \sum_{j=1}^{n} p(x_j|y_k) \log_2 [p(x_j|y_k)] \qquad (2)$$

where the conditional probability $p(x_j|y_k)$ describes how the probability that ion x_j is present has been affected by the result y of experiment k. The difference between the uncertainty at the start and the uncertainty after the experiment defines the amount of information gained in the test, namely,

$$I = H_{before} \sum_{k=1}^{m} p(y_k) H_{after}(k) \qquad (3)$$

Note that the uncertainty remaining after the experiment must be weighted by the probabilities of the m possible outcomes of the experiment. The quantity I is the Shannon information, which, because of the base 2 logarithm, is measured in binary digits, or *bits*.

For the qualitative analysis scheme, suppose that 10 ions are to be tested. One ion is present in the sample. For the purpose of this example, we presume that each test has one of two outcomes—a precipitate or no precipitate—and that these two results can be clearly determined. Further suppose that the addition of HCl produces a precipitate. The starting uncertainty, H_{before}, is calculated from Eq. (1) as $10[-0.1 \log_2 (0.1)] = 3.322$ bits. Based on the results of the experiment, the conditional probability $p(x|y)$ is 0.333 for lead, mercury(I), and silver, since we take each as equally likely to have produced the precipitate. The conditional probability is 0 for the other seven ions because we know that only one ion is present, and it must be one of the three ions that produce a precipitate with HCl. Thus, according to Eq. (2), H_{after} is $3[-0.333 \log_2 (0.333)] = 1.590$ bits. From Eq. (3), taking $p(y_k)$ as 1, since the signal "precipitate" can be discerned exactly, the information gain in this experiment is $I = 3.322 - 1.590 = 1.732$ bits. Had no precipitate formed, the conditional probability $p(x|y)$ for lead, silver, and mercury(I) would have been 0, while $p(x|y)$ would be 1/7 for the other seven ions (since we know nothing to aid in our identifying which of these remaining ones might be present). Thus, in this case, H_{after} is just $7[-0.1429 \log_2 (0.1429] = 2.813$ bits. Because the precipitate is certainly not present, $p(y_k) = 1$. From Eq. (3), the information gain is now $I = 3.322 - 2.813 = 0.509$ bit. In this case, the remaining uncertainty is considerably larger than when the precipitate formed, because fewer ions are known not to be present in the sample, but both results show less uncertainty than that present at the start. Thus, the precipitation test provides information, whatever the outcome, but some outcomes are more informative than others.

Problem 1. Suppose, for the 10-ion problem discussed above, that two ions were known to be present. The same HCl test is tried. Calculate the beginning uncertainty, the uncertainty remaining if a precipitate forms, and the uncertainty remaining if no precipitate forms. Discuss how you would generalize your results to the case where all 10 ions might be present.

To evaluate information content requires a knowledge of probabilities, however. First, the probabilities of the possible signals y_k must be known. This is not always simple, since the signal measured depends on the substance but can also be affected by other factors, including noise and experimental errors. Second, the probabilities of the identity of the unknown prior to analysis must be known. In essence, this forces the analyst to write down all preexisting information about the analysis. After all, the goal of a qualitative analysis is to make the probability for *one particular* x equal to 1 and the probabilities of all other x's equal to zero. When this situation is obtained, the uncertainty is zero. Intuitively, the largest uncertainty associated with a qualitative analysis arises when all members of a set are equally likely to be present. Finally, the conditional probabilities for relating analyte x to the *known* signal y, $p(x|y)$, must be known. In the example above, this relation was easily obtained because the test is simple, but often these conditional probabilities are especially difficult to obtain directly. It is more usual to know $p(y|x)$, the conditional probability of getting signal y from a known *analyte* x. Often, a range of signals y are possible in measuring a known analyte x and $p(x|y)$. The conditional probability can be calculated from more easily known quantities, however. Bayes' rule shows that

$$p(x_j|y_k) = \frac{p(x_j)p(y_k|x_j)}{\sum_{j=1}^{n} p(x_j)p(y_k|x_j)} \tag{4}$$

which permits ready calculation of $p(x|y)$ from the known or guessed prior probabilities of analytes $p(x)$ and the known relation of signal to analyte, $p(y|x)$.

The same equations apply to quantitative analysis. The variable x now refers to concentration, which is a discrete variable covering n possible values. In this case, however, it is not reasonable to attempt to obtain a probability of 1 for a *single* x, as is the goal of a qualitative analysis, because analytical precision is limited. Instead, the goal is to design experiments that maximize $p(x|y)$ for a small number ($<<n$) of values of x while minimizing $p(x|y)$ for other values. Thus, the result of a quantitative analysis is the sharpening of a distribution described by the values of $p(x)$ as a consequence of an analytical measurement. In this case the prior probability distribution might be uniform over some range, implying that no information was available on concentration, except for some upper and lower limits (which might be 0 and 100%). The conditional proba-

bility $p(x|y)$ will describe some distribution—perhaps even a Gaussian distribution. This final distribution arises because experimental error, noise, and random effects permit several possible values of x to be associated with the known (measured) signal y. If the analysis has been done well, this final distribution will have significant intensity in only a small fraction of the region spanned by the original uniform distribution. The narrowing of the distribution is reflected in a decreased uncertainty. Information is gained in the analysis.

Problem 2. Suppose a quantitative analysis of lead is desired. The sample may contain any amount of lead (0–100%). After 10 replicate analyses, a value of 0.135% lead is found, with a standard deviation of 0.011%. Calculate the information gained from the analysis.

Problem 3. Redundancy is defined as the difference between the maximum attainable value of the entropy, H_{max}, and the real value of the entropy. Redundancy represents the real amount of prior information available before the determination. The uniform distribution represents, as noted above, maximum uncertainty. Any restriction of the domain of the uniform distribution reduces overall uncertainty and produces redundancy. In the example above, suppose that it is known, prior to the analysis, that the lead sample concentration lies between 0 and 0.25%. Calculate the redundancy.

Redundancy also arises in changing from the uniform distribution to any other distribution. Suppose the lead content of the sample, which may contain from 0 to 100% lead, is believed to exhibit a normal distribution, with $\sigma = 16.66\%$. Calculate the redundancy.

Sometimes analytical measurements or tests are redundant. The total information gained is not the sum of the information gained from each single measurement or test. Redundancy here is defined as the maximum information gain minus the actual information gain. The example given above had no redundancy because only one test was done. Had we continued with the qualitative analysis scheme, the effects of redundancy would be felt, since some of the tests would give the same information as that gained on the first test.

Classification and Modeling as Information Problems

Now that the information gain has been calculated, of what use is it? Information content or gain cannot be used directly. The arbitrary definition of uncertainty and the unique units of uncertainty and information make raw information numbers useless. Instead, these quantities measure the efficiency of transmission of the important parts of data, so they are of use only in comparisons. The qualitative analysis example above shows a comparison of the information gained from having a precipitate form (a gain of about 1.7 bits of information, as a consequence of our being able to eliminate seven ions from consideration) as opposed to not getting a precipitate (where the gain is only about 0.5 bit, because only three of the 10 ions can be shown to be absent). The qualitative analysis is obviously a simple, and rather artificial, example, but it demonstrates the comparison well. In fact, the problem illustrated above can be seen to be identical to a classification of an object into one of several categories. In this example, we have presumed an unambiguous classification rule (a precipitate forms or it does not) and some separation of categories based on only one "feature" (the result from the HCl test). Qualitative analysis, then, is one form of pattern recognition, where past experience with cation chemistry has provided both the categories and the classification scheme. Not all such classifications are so thoroughly studied. Information gain from a classification can aid in identifying the most efficient process, and it can be of considerable use in selecting features and classification rules. Maximum dissimilarity of categories, as measured by information theory, has been used as a feature selection tool [3].

In the example concerned with quantitative analysis, the narrowing of distributions is reflected in a gain of information. Many chemometric methods produce, in one form or another, a distribution as one result of their application. For example the fitting of a line to a set of data may be regarded as the narrowing of a uniform distribution (if nothing is known prior to the fitting) to some final distribution that may reflect the random error in the data. In the ideal case, the fitting of a model "whitens" the data, extracting all systematic variations and leaving only random ("white") noise. Information gain is one way to monitor the efficiency of this process. Indeed, information is a good measure of optimization, of which fitting a model to data is just one form.

10.2 THE FUTURE OF CHEMOMETRICS

Simplification of Analytical Instrumentation

Over the past 30 years, the capabilities of analytical instrumentation have increased greatly, aided by advances in electronics and optics and better application of basic physics. As the instrumentation has improved, the tendency of some has been to simply increase the instrumental power directed at a problem in hopes that, with enough dimensions (wavelength, time, potential, and so forth), the solution will become obvious without the need to resort to much mathematical analysis. As instrumentation has become more complex, the informing power, a measure of the information available from an instrument [4], has increased significantly [5], but so has the redundancy. An alternative approach is to make use of simpler instrumentation with somewhat lower informing power (and usually less redundancy). Simpler and cheaper instrumentation makes good economic sense when used in monitoring chemical processes because more instrumentation can be purchased and more sites can be monitored, often in "real time." Chemometrics can be used to ensure that the information present in the response from these simpler, less informative instruments is extracted as completely as possible. This is the driving philosophy behind work in sensors today.

The chemometric methods required for the use of sensors as replacements for complex analytical instrumentation span a wide range. Calibration is very important, because the main task of the sensor is to convert some simple response, say a voltage from a surface acoustic wave (SAW) device, to an estimate of concentration of some species. The two quantities must be related. Then, there is a need to remove noise and other extraneous effects from the sensor response. Signal processing methods are essential here. Drift in the sensor must also be considered, often through use of time series analysis methods. And, of course, there is a need for work on materials capable of sensing the desired analytes in mixtures. These materials are conveniently identified by cluster analysis and classification methods [6].

Sensors can be used alone, but they are most useful in arrays, where several sensor "channels" provide information on the analytes in their detection zones. A sensor array can be built from discrete sensors such as the SAW device, but many relatively simple instruments are also conveniently thought of as sensor arrays. Near-infrared spectroscopy has been studied extensively as a relatively low cost sensor array for mixtures of organic species such as protein and starch [7]. These and other multichannel sensor arrays offer all the advantages of multivariate data for classification and calibration. The multivariate data also offer a special challenge to those concerned with data analysis, because the classical methods of signal analysis and time series analysis have not been developed for multivariate data. Thus, new methods for multivariate analysis must be developed and tested.

Sensors, single or multichannel, need not just "sense" concentration. For example, the infrared spectrum of a material may be related, through multivariate calibration, to the properties, such as thermal expansion or even taste, of the sample. Recent reports have shown that a number of diverse properties can be "sensed" through measurement of the near-infrared or infrared spectrum, including the energy content of natural gas [8] and the octane number of gasoline [9]. Calibrating a sensor array to a property of a sample is one way to avoid costly, difficult, and time-consuming tests. It also points out an advantage of a combined sensor–chemometrics approach over one based on hyphenated instrumentation: efficient extraction of *useful* information. In this case, the sensor is made specific to the desired property by the chemometric methods embedded with it. The hyphenated instrument may indeed have a much higher informing power, but, on the other hand, it may provide *no* information on sample properties if no relation between response and property is established. Finding ways to relate response to property is a new area of chemometric research that is likely to experience considerable growth in the next few years. Work in this area will involve development of better methods for calibration, study of error in calibration, and possibly the investigation of methods for calibration to clearly nonlinear relations. In addition to these empirical studies, investigation of the causal relationships that underlie response–property calibration is warranted. A related area, the correlation of chemical structure and property, is also likely to receive more attention.

Development of Intelligent Instrumentation

Not only can chemometrics be used to make simpler, cheaper, and more specific instruments, it can also be a key part of making "smarter" instruments. Calibration and modeling methods permit routine calculation of precision. Estimates of precision, or information as noted above, can be used to evaluate the quality of the calibration and modeling. With this information a variety of decisions can be made.

One application of quality monitoring is the simple tracking of the estimation process, to ensure that the estimation remains within some preset limits. A series of operations based on if-then logic permit action to be taken if the results fall outside the limits. For example, the estimated precision of a determination has been used with a recursive digital filter to monitor the quality of an automated determination in flow injection analysis. When the precision of the determination fell outside a preset threshold, the analyzer recalibrated until the precision was improved [10,11]. Self-calibrating instrumentation should have application in many areas, not just flow injection.

Control of processes is also possible. Current efforts at control by chemists are few, in part because chemists are more concerned with the monitoring of a process than with its stabilization over time. More often, chemical engineers have examined control theory and application. A considerable body of literature is now available. Control theory combines modeling, time series analysis, and calibration, all active areas of chemometrics research, so that ideas from chemometrics may be very useful. Research in control is increasingly directed to combined use of sensor arrays and chemometric methods, which range from simple if-then decisions to more sophisticated control methods based on optimal state estimation [12,13], fuzzy modeling [14], and other novel approaches.

Information gained from experiments may not always be purely mathematical. Heuristics and experience count as forms of information. Increasing use is being made of logical information in the control and optimization of experiments. Use of logical information is one of the fastest growing areas of chemometrics. Artificial intelligence (AI) is commonly applied in two situations:

where it is necessary to encode chemical expertise, and where the chemistry is merely another domain in which AI methods can be applied [15]. Whereas the trend in chemometrics research is to study the performance of expert systems in the former situation, many of the first applications of expert systems may be classified, understandably, into the latter situation. Knowledge-based (expert) systems [16] can apply the encoded chemical experience and expertise to solving a wide variety of problems, from optimizing the parameters of chemical separations [17] to classifying mass spectra [18]. In classifications, an expert system has performed as well or better than a more conventional classifier based on the SIMCA algorithm.

Intelligent instruments can also incorporate logical decision making into data reduction and control. Expert systems have been used to continuously optimize a triple-quadrupole mass spectrometer, to aid in collecting pyrolysis–gas chromatography–mass spectrometry data, and even to control the calibration of continuous analyzer systems, in conjunction with more mathematically oriented precision estimators, as discussed above. In fact, the combination of logical and mathematical methods would seem to offer a number of attractive features. Research is just now beginning in this area.

Adaptive Methods for Chemometrics: Fuzzy Methods and Neural Networks

If often happens that prior information, when available, is not without uncertainty itself. For example, the boundaries of categories in a classification problem may not be well separated, perhaps because there is no clear way to distinguish all members of the category. Using the common "hard" classification methods may lead to many misclassifications because the classification rule cannot account for the uncertainty, or "fuzziness," of the category boundaries. Building in an ability to handle fuzziness has become an important part of cluster analysis and classification, and it is of increasing interest in modeling. In developing methods that are more forgiving of imperfect prior knowledge, the goal remains the same as when more conventional methods were developed: extraction of the largest amount of useful information, given the quality of the input data and prior information [19]. Expert systems are especially suited to the treatment of fuzziness because prior knowledge and decision making can help sharpen the fuzziest of problems.

A particularly interesting form of fuzzy remodeling has only recently been explored. This method attempts to emulate, on a very superficial level, the action of the human brain in dealing with complex data such as spectra. First, the brain "learns" about the system from known, practice ("training") sets. Then, as in pattern recognition, it uses what it has learned to make conclusions about an unknown ("test") set of data. The "brain" here is a logic circuit known as a *neural network*. The neural network differs from conventional pattern classifiers in that is can adapt, in theory, to differing noise in the data, and it permits a certain amount of fuzziness in the classification. Both are important advantages, since noise and pattern uncertainty often limit classifications. The method for learning and prediction is essentially an adaptive form of pattern recognition, with roots in fuzzy classification and in expert systems. A neural network application has only recently been published [20], but many more should appear soon.

SOLUTIONS TO PROBLEMS

1. The beginning uncertainty is calculated from Eq. (1), assuming that any pair of the 10 ions is equally likely. The chance of drawing two ions from a set of 10 is given by the combinatorial rule $^nC_k = n! / k! (n-k)!$, the number of ways the sequence containing k outcomes 1 and $n-k$ outcomes 2 can be generated, and where $n! = n(n-1)(n-2)(n-3) \ldots (n-(n-1))(1)$.

Here, $k=2$ and $n=10$, so that $^nC_k = 10!/2! \, 8! = 45$ possible combinations of ions. Since each combination is equally likely, $H_{\text{before}} = \log_2(45) = 5.49$ bits.

Formation of the HCl precipitate indicates that either one or both of the ions are lead, silver, or mercury(I). Therefore, no more than one of the other ions is present. The number of combinations that contain at least one lead, silver, or mercury(I) is $7+7+7+3 = 24$. Thus the uncertainty is $\log_2(24) = 4.59$, a gain of only 0.90 bit. When no precipitate forms, all that is known is that neither ion that is present is one of the above three ions. The number of possibilities that remain is ^nC_k with $n=7$ and $k=2$, or 21 combinations. The uncertainty is $\log_2(21) = 4.39$ bits. The information gain is 1.10 bits, more than in the one-ion case because the additional uncertainty caused by having two ions makes a negative HCl precipitation result more useful than a positive result.

If all 10 ions might be present, we could have any combination containing from all 10 ions to one ion. The number of combinations that are possible can be figured from the combinatorial rule by considering the number of combinations from mixtures of 1, 2, 3, . . . , and 10 ions. Summing the total number of combinations gives the starting uncertainty, which will be very large.

A precipitate indicates that *at least* one to three ions are present, and others may also be present. Thus, any of the combinations containing at least one of the test ions is possible. These can be calculated by the combinatorial rule, as before. No precipitate indicates that any combination with any of the three ions is *not* present, but any combination containing one or more of the others is still possible, and the number of these can be determined by the difference of the number of combinations with precipitating ions from the total number of combinations.

2. The initial uncertainty is calculated on the basis of a uniform distribution from 0 to 100%. The smallest value that can be distinguished, Δc, 1%, judging from significant figures. Thus, $H_{before} = \log_2(100) = 6.64$ bits. Replicate analyses reduce the uncertainty to a Gaussian distribution, centered at 0.135%, with standard deviation of 0.001%. Evidently, $\Delta c = 0.001$ now. For this distribution, the entropy is $H_{after} = \log_2 \sigma \sqrt{2\pi e}/\Delta c = 5.51$. Thus, the information gain is 1.13 bits.
3. The additional information reduces the uncertainty from the uniform distribution 0–100% to the uniform distribution 0–0.25%, for which $H_{before} = 4.64$ bits. The redundancy is therefore $6.64 - 4.64 = 2.00$ bits. In the second situation, the additional information reduces the uncertainty from a uniform distribution 0–100% to a Gaussian distribution with $\sigma = 16.66\%$, where (taking Δc as 1%) we find $H_{before} = 6.11$ bits. The redundancy is therefore $6.64 - 6.11 = 0.53$ bit.

REFERENCES

1. E. Shannon and W. Weaver, *The Mathematical Theory of Communication*, University of Illinois Press, Urbana, Ill. (1949).
2. K. Eckschlager and V. Stepanek, *Information Theory as Applied to Chemical Analysis*, Wiley, New York. (1979).
3. K. Varmuza, *Pattern Recognition in Chemistry*, Springer-Verlag, Berlin, Chapter 11. (1980).
4. H. Kaiser, *Anal. Chem.*, 42(2):24A (1970); 42(3):26A (1970).
5. J. D. Winefordner, ed., *Trace Analysis*, Wiley, New York, Appendix A. (1976).
6. D. S. Ballantine, Jr. and H. Wohltjen, *Anal. Chem.*, 61:704A (1989).
7. E. Stark, K. Luchter, and M. Margoshes, *Appl. Spectrosc. Rev.*, 22:335 (1986).
8. S. M. Donahue, C. W. Brown, B. Caputo, and M. D. Modell, *Anal. Chem.*, 60:1873 (1988).
9. J. J. Kelly, C. H. Barlow, T. M. Jinguji, and J. B. Callis, *Anal. Chem.*, 61:313 (1989).
10. P. C. Thijssen, S. M. Wolfrum, G. Kateeman, and H. C. Smit, *Anal. Chim. Acta*, 156:87 (1984).
11. P. C. Thijssen, G. Kateman, and H. C. Smit, *Anal. Chim. Acta*, 157:99 (1984).
12. T. Chattaway and G. N. Stephanopoulos, *Biotechnol. Bioeng.*, 34:647 (1989).
13. J. L. Harmon, G. Lyberatos, and S. Svoronos, *Biotechnol. Bioeng.*, 33:1419 (1989).
14. K. Konstantinov and T. Yoshida, *Biotechnol. Bioeng.*, 33:1145 (1989).
15. N. A. B. Gray, *Anal. Chim. Acta*, 210:9 (1988).
16. J. W. Frazier, *Mikrochim. Acta*, 2(1–6):163 (1986).

17. D. Goulder, T. Blaffert, A. Biorkland, L. Buydens, A. Chabra, A. Cleland, N. Dunand, H. Hindriks, and G. Kateman, *Chromatographia*, 26:237 (1988).
18. D. R. Scott, *Anal. Chim. Acta*, 223:105 (1989).
19. M. Otto, *Chemometr. Intell. Lab. Syst.*, 4:101 (1988).
20. J. U. Thompsen and B. Meyer, *J. Magn. Resonance*, 84:212 (1989).

FURTHER READING

Litenu, C., and I. Rica, *Statistical Theory and Methodology of Trace Analysis*, Ellis Horwood, Chichester. (1980).

Eckschlager, K., and V. Stepanek, *Analytical Measurement and Information: Advances in the Information Theoretic Approach to Chemical Analysis*, Research Studies Press, Letchworth, Hertfordshire, U.K. (1985).

Martens, H., and T. Naes, *Multivariate Calibration*, Wiley, Chichester, U.K. (1989).

Pao, Y.-H., *Adaptive Pattern Recognition and Neural Networks*, Addison-Wesley, Reading, Mass. (1989).

Pierce, T. H., and B. A. Hohne, *Artificial Intelligence Applications in Chemistry*, ACS Symp. Ser. 306, ACS, Washington, D.C. (1986).

Index

Accuracy, 14–17
Algorithm design, 285–286
  bottlenecks, 286
  concurrency, 285
  parallelism, 285
  vectoring, 285
Analysis of variance (ANOVA), 26–32, 57, 184
  difference between means, 26
  within sample variation, 27
  between sample variation, 28
Artificial intelligence, 305–307

Bias, 14–15
  laboratory, 15
  method, 15

Calibration, 99–149
Central limit theorem, 10, 103
Central processing unit, 275, 281
  communications, 277
  input/output, 277

Chi squared test, 34–36
Cluster analysis, 216
  adaptive least squares, 222
  classification, 219–227
  dendrogram, 217
  linear discrimination methods, 221
Coefficient of variation, 11
Computer hardware, 271–280
  in-line, 273
  memory, 275
  on-line, 272
Computer interfacing in the laboratory, 278, 292–303
  analog interfaces, 292
  BUS, 279
  data storage, 278
  IEEE, 295
  parallel digital interface, 298
  real-time operations, 299
  RS232, 297

321

Computer interfacing (*continued*)
  software, 280
  standard interface, 294
Computer performance characteristics, 280–283
  accelerators, 281
  floating point, 282
  speed, 281
  word size, 280
Confidence intervals, 124–128
  calibration, 125
  intercept and slope, 124
Confidence level, 16–18
Confidence limits, 32, 53
Convergence methods, 154–162
  error surface, 156
  modified simplex method, 155
  normalized residual plot, 159
  unweighted error sum, 154
  weighting factors, 156

Data analysis, 213–236
Deconvolution, 262
Degrees of freedom, 11
Dendrogram (*see* cluster analysis)
Derivative methods, 264
Dixon Q-test, 33

Eigenvalues, 215
Eigenvectors, 215
Error, 7–16
  random, 8
  systematic, 14
Euclidean distance, 215
Experimental design, 181–198
  comparative experiments, 183
  factional factorials, 197
  factorial design, 189–198
  factorial experiments, 183
  interactions, 190
  latin squares, 187
  randomized blocks, 184
  2^n factorial designs, 190
Expert systems, 305
Exploratory data analysis, 42, 62–79
  box and whisker plots, 70
  letter values, 66, 73
  resistant regression, 76
  scatterplots, 62
  skewness in data, 74
  stem and leaf displays, 64
Exploratory data analysis multivariate, 79–85
  Andrews curve, 81
  Chernoff faces, 79
  Kleiner-Hartigantrees star plots, 83
  two-way ANOVA, 83

F-test, 16, 19–22
Factorial design (*see* Experimental design)
Filtering in the frequency domain, 242–255
  auto correlation, 249
  discrete Fourier transform, 245
  fast Fourier transform, 246
  Fourier transform, 242
  Nyquist frequency, 243
  smoothing, 251
Filtering the time domain, 255–262
  Savitsky-Golav, 260
  smoothing, 255
Fourier transform (*see* Filtering in the frequency domain)
Fuzzy methods, 317

Gaussian distribution, 8, 39, 101
Goodness of fit (*see* Model response curves)

Index

Information theory, 310–314
Intelligent instruments, 316
Interferences, 137–138

Kolmogorov-Smirnov, 58
Kruskol-Wallis, 58

Laboratory information
    management systems
    (LIMS), 303–305
Laboratory interfacing (see
    Computer interfacing in the
    laboratory)
Least squares technique (see
    Regression analysis)
Limits of detection, 132–137
Linear calibration, 111–128
    alternative multicomponent
        analysis, 118
    collinearity, 123
    curvilinear regression, 113
    least squares, 111
    multivariate regression, 115–124
    regression diagnostics, 120
    regression through the origin,
        113
    univariate regression, 112
Linear and nonlinear models, 152
Linear regression, 153

Mann-Whitney-Wilcoxon test, 55
Mapping techniques (see Pattern
    recognition)
Mean, 8
Median, 50
Microcomputing, 283–303
    connectivity, 287
    performance, 283
    transputer, 284
Microprocessor, 271–307
Model response curves, 162–168
    closed form, 162

empirical models, 164
goodness of fit, 168
solvable equations, 166
Monte Carlo simulations, 223
Multivariate exploratory data
    analysis (see Exploratory
    data analysis multivariate)
Multivariate regression (see Linear
    calibration)

Neural networks, 306, 317
Non-linear calibration, 128–130
Non-linear regression analysis,
    151–179
    applications, 168–179
Non-parametric methods, 49–98
Normal distribution, 8, 34, 39, 101
Null hypothesis, 16, 104

Optimization, 198–209
    multi factor, 201
    response surfaces, 198
    Simplex, 205
    single factor, 199
    steepest gradient techniques, 202
Outliers, 32–34

Parametric methods, 5–38
    limitations, 43
Partial least squares, 119
Pattern recognition, 211–236
    applications, 227–236
    data processing, 213
    data representation, 212
    mapping and display, 213
    non-linear mapping, 215
Precision, 8–14
Principal component analysis,
    213–216
    eigenvalues, 215
    eigenvectors, 215
    euclidean distance, 215

Principal component regression, 118
Programming languages, 288–292
    Assembly language, 288
    BASIC, 289
    C, 289
    debugging, 291
    FORTH, 290
    FORTRAN, 288
    Pascal, 289

Regression analysis, 36–37, 43
    least square technique, 36
    ordinary least squares, 45, 60
Relative standard deviation, 11
Robust statistics, 40

Sampling theorem (see Sampling theory)
Sampling theory, 100, 243
    distributors (see Normal distribution)
    fixed sampling plan, 105
    sequential sampling plan, 107
Savitsky-Golav (see Filtering in the time domain)
Selectivity, 138–140
Sensitivity, 130–132

Signal processing, 211, 213, 239, 269
    noise removal, 240
Significance testing, 15–26
SIMCA, 225–227
    cross validation, 286
Simplex (see Optimization)
Smoothing (see Filtering in the frequency domain)
Software, 286–288
    operating system, 286
Standard additions, 140–146
    multivariate, 145–146
    univariate, 140–145
Standard deviation, 9–14
Standard error of the mean, 12
Statistical methods, 5–38

t-distribution, 17, 103
t-test, 15–25
    comparison of means, 23
    estimation of accuracy, 22
    paired test, 24
    type I and type II error, 25, 104

Variance, 11

Wilcoxon signal rank, test, 50